The Field Naturalists Club of Victoria

The Victorian Naturalist

The Field Naturalists Club of Victoria

The Victorian Naturalist

ISBN/EAN: 9783743311800

Manufactured in Europe, USA, Canada, Australia, Japa

Cover: Foto ©berggeist007 / pixelio.de

Manufactured and distributed by brebook publishing software
(www.brebook.com)

The Field Naturalists Club of Victoria

The Victorian Naturalist

The
Victorian
Naturalist

Volume 130 (1)   February 2013

Published by The Field Naturalists Club of Victoria *since 1884*

From the Editors

This year began with a January that was not excessively hot, and while February has indeed been warm with temperatures consistently rising into the low thirties, the over-arching weather feature in Victoria has been the lack of rain. This summer 2012/13 must be one of the driest on record. We have had fires, both deliberately lit and naturally sparked, which have caused widespread havoc in some of our more natural bushland areas. But the bush is remarkable for its ability to renew itself after fires through regeneration. As long as fires occur infrequently we can expect to maintain the biodiversity of these areas.

At *The Victorian Naturalist* our year begins with a substantial issue of the Journal containing papers on a wide spectrum of subjects with species from the Animal Kingdom, both vertebrate and invertebrate, strongly represented. Also we pay tribute to and farewell Dorothy Mahler, a faithful servant of the Club.

The Victorian Naturalist
is published six times per year by the

Field Naturalists Club of Victoria Inc
Registered Office: FNCV, 1 Gardenia Street, Blackburn, Victoria 3130, Australia.
Postal Address: FNCV, Locked Bag 3, Blackburn, Victoria 3130, Australia.
Phone/Fax (03) 9877 9860; International Phone/Fax 61 3 9877 9860.
email: admin@fncv.org.au
www.fncv.org.au

Patron: His Excellency, the Governor of Victoria

Address correspondence to:
The Editors, *The Victorian Naturalist*, Locked Bag 3, Blackburn, Victoria, Australia 3130.
Phone: (03) 9877 9860. Email: vicnat@fncv.org.au

Yearly Subscription Rates – The Field Naturalists Club of Victoria Inc
(As of October 2012)

Membership category		Institutional	
Single	$75	Libraries and Institutions	
Concessional (pensioner/Senior)	$55	- within Australia	$140
Additional Concessional	$20	- overseas	AUD150
Family (at same address)	$95		
Junior	$25		
Additional junior (same family)	$10	Schools/Clubs	$85
Student	$30		

All subscription enquiries should be sent to
FNCV, Locked Bag 3, Blackburn, Victoria, Australia 3130.
Phone/Fax 61 3 9877 9860. Email: admin@fncv.org.au

The
Victorian
Naturalist

Volume 130 (1) 2013

February

Editors: Anne Morton, Gary Presland, Maria Gibson

Editorial Assistant: Virgil Hubregtse

ISSN 0042-5184

Front cover: *Lipotriches australica* sleeping aggregation, December 2006. Photo by P Kubiak.
Back cover: Pacific Black Duck *Anas superciliosa*. Photo by Anne Morton

Research Reports

Birds of Seal Rocks in northern Bass Strait over 40 years (1965–2005)

Robert M. Warneke[1] and Peter Dann[2]

[1]Blackwood Lodge, 1511 Mt Hicks Road, Wynyard, Tasmania 7325.
[2]Research Department, Phillip Island Nature Parks, P. O. Box 97, Cowes, Phillip Island, Victoria 3922
E-mail: pdann@penguins.org.au

Abstract

Long-term datasets of fauna are rare for uninhabited islands in south-eastern Australia. Here we report on 40 years of observations from 1965 to 2005 on the birds of Seal Rocks in northern Bass Strait. Seventy-five native and six exotic species including 21 native passerines were observed at Seal Rocks or nearby. Six species were recorded breeding—Crested Tern *Thalasseus bergii*, Silver Gull *Chroicocephalus novaehollandiae*, Sooty Oystercatcher *Haematopus fuliginosus*, Welcome Swallow *Hirundo neoxena*, Common Starling *Sturnus vulgaris* and, for the first time in Victoria, Kelp Gull *Larus dominicanus*. The main changes to breeding birds over the 40 years have been the movement and expansion of the breeding colonies of Crested Terns and Silver Gulls to adjacent parts of nearby Phillip Island, and the arrival and expansion of the breeding Kelp Gull population. Kelp Gulls have increased substantially at Seal Rocks since their arrival in 1968. The first reported breeding for Victoria occurred there in 1971. The expansion of Kelp Gulls may have been associated with the expansion of Australian Fur Seal *Arctocephalus pusillus doriferus* numbers which, on one hand, has reduced the number of suitable breeding sites but, on the other hand, has increased the amount of food available in the forms of vomited food remains and placentae. The variety of land birds recorded on Seal Rocks was surprisingly high, given the exposed nature and relative sterility of the terrain; however, the strait between Seal Rock and Phillip Island is narrow and all the species recorded there are common in the region and most are wide-ranging seasonally or undertake significant north-south migrations. Records were relatively few after 1997 due, in part, to the reduced amount of vegetation on the islets and greatly reduced lengths of research stays. (*The Victorian Naturalist* 130 (1) 2013, 4–21).

Key words: Seal Rocks, Phillip Island, birds, long-term survey

Introduction

Seal Rocks (38° 32' S, 145° 06' E), a State Faunal Reserve, comprises two small islets—Seal Rock and Black Rock—which lie about 1.5 km off the south-west point of Phillip Island, Victoria (Fig. 1). Seal Rock, the larger islet, includes two detached rocks to the north-west and an extensive area of low-lying reef to the east, which is cut off at high tide (Fig. 2). The total land area is about 2.8 ha.

Although Seal Rocks has had a long history of visits by Europeans dating from 1801 (Warneke 1982, 2003), very few detailed accounts survive and none provide any useful information on the vegetation or bird life. This paper reports on incidental sightings and observations on birds ashore and at sea around the islets, accumulated during a program of research on the resident colony of Australian fur seals *Arctocephalus pusillus doriferus*, by Robert Warneke (1965–1991) and continuing studies there on fur seals and birds by Roger Kirkwood (RK) and Peter Dann (1998–2005). Datasets of such duration are rare, particularly for uninhabited islands in south-eastern Australia.

Field studies were initiated in 1965 by the Fisheries and Wildlife Department (now the Department of Sustainability and Environment) and supported until 1979. A field station was built on the southern plateau of Seal Rock in the summer of 1965–1966 (Warneke 1966) and subsequently other structures were erected to facilitate aspects of the project including a flying-fox connecting the two islets (1967), a small observation hide on each of the Seal Rock plateaux (1967) and a 15 m steel observation tower adjacent to the station (1969). Most of this infrastructure was demolished by 1979 when a second field station was built out of rock at the base of the northern end of South Plateau.

A small research team visited Seal Rocks at approximately monthly intervals from 1966 until 1972, but thereafter visits were limited to November and December each year to monitor events during the breeding season and to count pups. Pups were routinely marked each January until 1977. From 1979 to 1991 RMW continued the November–December monitor-

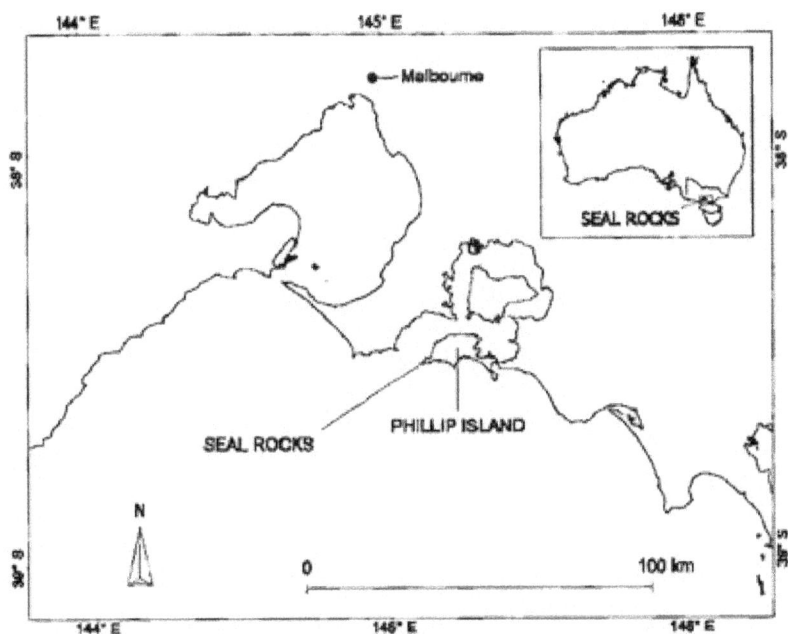

Fig. 1. Locations of Seal Rocks and Phillip Island in Bass Strait, south-eastern Australia

ing and counting on a private basis until, by 1991, very few marked animals remained alive (Warneke 2003). More recently, Seal Rock has been visited on 40 occasions between February 1997 and April 2005 by PD and RK. Most visits were of several hours' duration but several trips lasted 3–4 days while satellite transmitters were deployed on fur seals (Kirkwood et al. 2005).

Opportunities for observation were constrained by the demands of the seal study, and although 'new' or rarely seen species and significant activities were always duly noted, observations on resident species were not systematically recorded. Visits to many parts of Seal Rock were consciously limited, to avoid disturbing the seals, and extra care was taken when birds were nesting; visits to check the contents of particular nests were cautious and brief. Landings on Black Rock were difficult and risky until the flying fox was in place, but even then that islet was visited only for specific tasks such as tagging seal pups.

The data accumulated under these circumstances, especially during the most intense period of seal research from 1966 to 1972, provide a fairly clear picture of bird diversity on the islets, including the occurrence of seasonal migrants. Most observations were made with the aid of Zeiss 8 x 30 binoculars, and frequently an 800 mm telescope was used for closer viewing of a particular bird or activity. The effective visual range was usually about 1 km for large seabirds or congregations, and as far as 3 km in clear weather using the telescope. Observational effort differed between the period of intense seal research (1966-1972) and the later years (1997-2005). These are referred to hereafter as the first and second survey periods respectively. Bird observations in the later years were confined to the islets and waters immediately adjacent, using 10 x 20 binoculars; consequently, many fewer pelagic seabirds were recorded.

Records of birds summarised here consist of notes in RMW's field journals, to which all

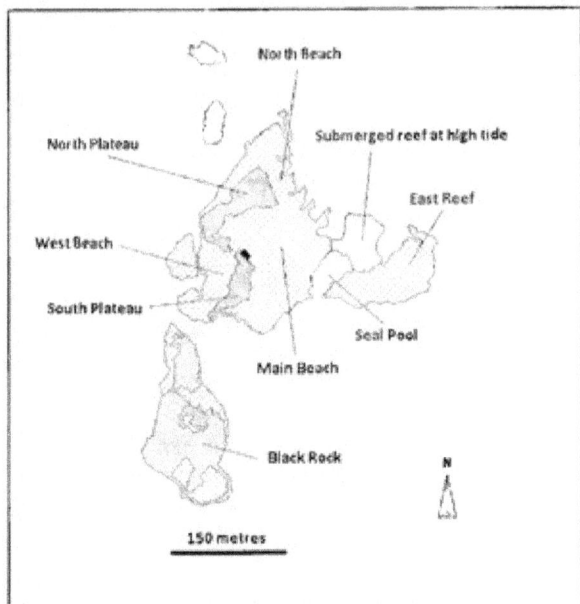

Fig. 2. Names used for locations on Seal Rocks.

team members' contributed sightings; brief entries in surviving personal diaries kept by Fred Baum (1965-1968, 1970-1971) and Kevin Chipperfield (1968), which include visits when RMW was not present; and Phillip Island Nature Park research group's records of 40 visits between 1997 to 2005. A negative linear binomial generalised model was used for looking at the seasonal pattern of Kelp Gull numbers using R software (version 2.15.1; R Development Core Team 2012).

Physical and biological features of Seal Rocks
The continuing effects of tides and storms on the geology of the islets have resulted in a varied topography of plateau tops, cliffs with ledges and undercuts, caves, gullies, boulder-strewn upper slopes, cobblestone-pebble shingle, worn wave-washed shore platforms with some deep tide pools, and many ephemeral tide pools flushed only by the highest tides or storm seas. Both islets are basically low platforms of fine-grained black basalt surrounding areas of plateaux of varying height to about 12 m above sea level. These plateaux are remnants of later volcanic events and consist of deposits of tuff overlain by lava flows. Storm waves have eroded the softer tuff and this has led to undercutting and rock falls from above. Many cliff faces are sheer with narrow ledges, and shallow caves occur at the south end of South Plateau and on all sides of Black Rock's central plateau.

Land plants survive only where the seals have no or only limited access, i.e. on cliff faces and some parts of the plateaux tops. Gravel from weathering of the volcanic rock tends to accumulate on ledges and in fissures on cliff faces, and supports clumps and mats of Rounded Leaf Noon Flower *Disphyma australe*. Along the margins of abrupt cliff tops, where resting seals are less inclined to lie, this plant has a precarious hold and generally forms a narrow ridge-like mat. Seedlings of Ruby Saltbush *Enchylaena tomentosa* were occasionally found, apparently imported via bird faeces or pellets. If they germinated in protected places on the cliffs they flourished for several years. Other colonists were Bower Spinach *Tetragonia implexicoma*, Cape Weed *Cryptostemma calendula*, Sow

Thistle *Sonchus* sp., a nightshade *Solanum* sp. and Boxthorn *Lycium ferrosissimum*, but, with the exception of Bower Spinach and Southern Sea-heath *Frankenia pauciflora*, none survived for long. Small mats of Beaded Glasswort *Salicornia quinqueflora* were found on Black Rock plateau in 1968 and 1971. A profusion of marine plants and invertebrates occurred at the edges of the shore platforms and on East Reef in the intertidal zone, including seaweeds, kelp *Macrocystis* sp., cunjevoi *Pyura* sp., barnacles, limpets, chitons, mussels (mainly *Xenostrobus* sp.) and the gastropods *Subninella* sp. and *Nerita* sp. were very common. Dense beds of kelp also occurred in the shallow bay formed by the two islets on the eastern side.

The insect fauna appeared to be diverse and most forms were seen only during the warmer months. It included field crickets *Teleogryllus commodus*, at least six different kinds of wasps (especially the orange ichneumon *Netelia* sp.), a black ant and a small black native bee; blowflies (*Calliphora* sp.) and a small black fly common in humid weather and attracted to sweaty skin; dragonflies and damselflies; moths (most notably *Agrostis* sp.) and butterflies (including *Vanessa kershawi*), and mosquitoes, noticeable only when a sheltered depression in the lee of the south end of North Plateau was sporadically filled by sea spray and rain showers, in which larvae were observed.

Species notes
Nomenclature and systematic order follows Christidis and Boles (2008).

Stubble Quail *Coturnix pectoralis*
Eight records of lone birds sighted on or about South Plateau, among the mats of noon flower or in the cover of rocks, in November of 1965, 1967 and 1969, December 1971, and in October and December 1975. The desiccated remains of a bird were found on South Plateau on 16 November 1967. Specimen RW#465, collected 25 November 1969 (Museum Victoria).

Black Swan *Cygnus atratus*
Four records of birds in transit; eight were seen heading west on 15 December 1969, one heading south on 19 December 1970, 11 heading west on 23 February 1972, and five heading north on 18 November 1975.

Australian Shelduck *Tadorna tadornoides*
A line of seven passed by to the north on 20 December 1974 flying towards the west.

Pacific Black Duck *Anas superciliosa*
A pair on 19 October 1970 flying slightly east of south, but they turned west before being lost to view.

Rock Dove *Columba livia*
Seven sightings of lone birds during the first survey period, in January and between May and August, and one of a group of three in November. Most were seen flying over or past the islets and generally in a northerly direction. On two occasions, lone birds landed.

Common Bronzewing *Phaps chalcoptera*
One record only, of a bird seen to pitch into the noon flower sward at the north end of South Plateau on 24 April 1968. It appeared to be moulting, having only one remaining loose tail feather.

White-throated Needletail *Hirundapus caudacutus*
One sighting of a lone bird that passed low over Seal Rock from the north-east at 17.30 hr on 17 January 1969 shortly after a north-westerly change. This species has been sighted over tidal flats on Western Port (Davies and Reid 1975b).

White-faced Storm-Petrel *Pelagodroma marina*
One record of a bird that flew into the field station through the open door on 18 October 1967 at 20.00 hr.

Wandering Albatross *Diomedea exulans*
Twenty-two sightings offshore during the first survey period, between mid-June to mid-November, and two additional records in mid-January. Most sightings were of lone birds gliding in rough windy weather, well offshore within a broad arc from the south-east to the north-west, and predominantly during south-westerly blows. Six sightings were of lone birds flying in calm conditions.

Black-browed Albatross *Thalassarche melanophris*
The most commonly observed albatross, from April to mid-December, but the majority of sightings (247 of 262) were from June to August; frequently in association with Shy Albatross *Thalassarche cauta*. Peak numbers were

recorded in the latter parts of May or June, usually on days of very rough weather. The greatest concentration was noted on 23 June 1971 during a hard south-westerly gale, when over 200 were in view within an arc from the south-east to the south-west. This species was observed only once in the second survey period when ten birds were seen on 25 May 1999. Despite frequent sightings close to shore early in the first survey period, feeding was rarely observed. On two occasions three to six birds fought over flotsam, identified in one instance as a large cuttlefish (probably *Sepia apama*); on another, two birds were observed harassing a seal thrashing a moderately large prey item at the surface, and on a third occasion several albatross paddled up to an Australasian Gannet *Morus serrator* when it surfaced with a fish and attempted to snatch it.

Shy Albatross *Thalassarche cauta*

Active offshore in small numbers from late April to mid-December (28 records), but with no obvious peak or influx as in the case of the preceding species. Possible feeding activity was observed on 15 June 1969 when several birds settled on a patch of discoloured water, which may have been a surface shoal of fish, to the south of Black Rock.

Southern Giant-Petrel *Macronectes giganteus*

Giant-Petrels were frequently seen offshore from June to December throughout the 40 year period, with a few additional sightings of lone birds in January, February, April and May. Dark-plumaged individuals predominated, with only three all-white (June, September), one grey (June) and one pale-headed individual (June) in 157 sightings. Only one record was made in the second survey period (June 2005).

Despite Giant-Petrels being avid scavengers of dead seals on many subantarctic seal islands and the frequent presence of seal carcases on Seal Rocks, only two instances of scavenging by Southern Giant-Petrels were observed, on 27 May 1972 and 24 June 1972. However, Giant-Petrels were seen feeding on carcasses floating offshore on several occasions, and once two Giant-Petrels were observed paddling about over a concentration of Coastal Krill *Nyctiphanes australis* and small fish, which in turn had attracted albatrosses, gannets and other small seabirds.

On three occasions a lone Giant-Petrel was seen ashore on the Main Beach breeding area, resting or walking among the fur seals with wings partly opened. Juvenile seals and even adult males retreated, whereas cows with young pups responded with open-mouthed threats.

Northern Giant-Petrel *Macronectes halli*

Only one certain record of a bird resting on the water off East Reef on 5 February 1979 (RMW, FTB).

Southern Fulmar *Fulmarus glacialoides*

A lone bird was sighted by KJC flying low over North Beach at midday on 22 August 1968.

Cape Petrel *Daption capense*

Eight sightings of lone birds between June and September, of a pair on 22 August 1968, and of a lone bird in mid-December. Usually seen flying low over the water and if alighting only for brief periods. On two occasions a Cape Petrel hovered about in the vicinity of a resting Giant-Petrel, but no interaction occurred. Specimen RW#216 collected 22 August 1968 (Museum Victoria).

Fairy Prion *Pachyptila turtur*

Observed once, on 16 June 1969. A group of 20-30 dived repeatedly into a swarm of krill located about 1.5 km south of Black Rock in company with feeding Short-tailed and Fluttering Shearwaters and White-fronted Terns. Specimen RW#414, collected 16 June 1969 (Museum of Victoria), a male, testes minute, heavy sub-cutaneous fat deposits, stomach filled with Coastal Krill *N. australis*.

Short-tailed Shearwater *Ardenna tenuirostris*

The daily passage of large numbers of this shearwater to and from breeding colonies on the south-west and south coasts of Phillip Island was a feature of the offshore bird activity from October to February each year. Thereafter to the beginning of May the number and regularity of sightings declined. Observations fell into three broad categories — morning exodus and evening return, localised activity at the sea surface within 3-4 km of Seal Rocks, and individual birds at or close inshore.

The morning exodus in calm to moderate weather was an orderly stream passing East

Reef heading to the south and south-west; no movement to the south-east was observed. In strong winds and heavy seas the birds were active offshore all day, widely dispersed on all sides and seemed to circle the islets both clock-wise and counter-clockwise. Landward move-ment in the evening was always diffuse. In flat calm seas, large rafts (c. 300-500 individuals) of these shearwaters were occasionally noted about 3-4 km offshore to the south-east and always in the same general vicinity. During the morning these rafts continually broke up and reformed nearby, but if present in late after-noon they were more stable, with most birds resting quietly. Infrequently, large flocks and small groups were seen feeding at the surface to the south-east, south and south-west. Tempo-rary rafts formed when birds alighted over con-centrations of krill, dipping and diving beneath the surface. In January and February 2000, c. 8500 and 200 respectively were recorded feed-ng within several kilometres of the islets.

Crested Terns and Fluttering Shearwaters were occasionally seen feeding with Short-tailed Shearwaters on the same concentrations. Most unexpectedly, a few Short-tailed Shear-waters were seen in June 1969, on several oc-casions during the 15th, 18th and 19th. The context of these sightings was remarkable in that these birds were active on a large swarm of Coastal Krill, together with Fairy Prions, Flut-tering Shearwaters, White-fronted Terns, Silver Gulls, Black-browed and Shy Albatross, Giant Petrels and Australasian Gannets. However, only the shearwaters, prions and terns were feeding directly on the crustaceans.

On rare occasions lone Short-tailed Shearwa-ter were seen resting on the sea, and in calm conditions were prone to attack by predators. In one instance two immature Pacific Gulls repeat-edly buffeted the Shearwater when it attempted to lift off the surface, knocking it down into the water. On another occasion two Giant-Petrels were seen tearing at a helpless bird floating at the surface. Despite opportunities for Austral-ian fur seals to prey on Short-tailed Shearwater in nearby waters, no instances were observed. Similarly, Deagle et al. (2009) found no evi-dence of shearwaters (or any other birds) in the faeces of fur seals at Seal Rocks.

Fluttering Shearwater *Puffinus gavia*
Recorded on 16 occasions offshore in the first survey period, most frequently in June-July and October-November, usually in groups of 2-5 birds. On 3 November 1969 at least 50 were observed feeding within 100 m of North Beach. Sightings in January, April and Septem-ber were of single birds. In calm weather feed-ing birds rose from the surface, flew a short distance and plunged in with wings extended. In rough weather they would fly into the face of an oncoming wave, emerge from the rear and fly into the next. By dropping back 50 m or so with the wind after passing through a succes-sion of waves in the same general vicinity would be worked in this way for about an hour. Speci-mens RW#116, 16 June 1969; RW#449, 20 Oc-tober 1969 (Museum Victoria).

Common Diving-Petrel *Pelecanoides urina-trix*
One record only, of a desiccated carcass found in the enclosed South Plateau observation hide on 26 May 1976. It had entered via a narrow gap in the roof at some time after 7 January 1976. On 25 May 1976 RMW, on board the *Lorraine May* of San Remo, observed a widely dispersed group of 50+ Diving-Petrels off Pyramid rock, approximately 13 km east of Seal Rocks.

Little Penguin *Eudyptula minor*
Breeding was not observed, but small numbers came ashore to moult or rest or because they were sick or injured. Moulting birds were found from mid-January to mid-April, but as sites free of disturbance by seals were few, the maximum number of birds found at any one time was eight. Moulting birds hid behind fallen boul-ders at the base of South Plateau and North Plateau or in shallow caves, or took advantage of artificial shelter afforded by a wood stack and a section of flooring stored under a cliff over-hang. Four single birds were found moulting in crevices around South Plateau between Febru-ary and May 1997-2005.

Penguins found at other times of the year (24 alive, 21 dead) were all in light to poor condi-tion; in eight instances they were wholly or partly stained by oil. A young lightly oiled bird found on 16 January 1973 was infested with a large number of ticks attached to the back of its head and neck. Ten carcasses were found within tide reach and may have died at sea, but the remainder had died where they had taken shelter. On 10 occasions, live penguins were seen during the day among fur seals resting on Main Beach and North Beach, and were completely ignored by the seals, although the Cape Fur Seal *A. pusillus pusillus* and the New Zealand Fur Seal *A. forsteri* are known to prey upon penguins (Shaughnessy 1978; Page *et al.* 2005). Although in plain view of Pacific and Silver gulls, on only one occasion was a penguin harassed and forced to retreat to the sea. Penguins were occasionally seen feeding within about 100 m of the shore and often were heard calling in calm, foggy weather.

Frigatebird *Fregata* sp.
A single sighting, on 20 December 1968 by RMW and WMB. Shortly after 13.00 hr, attention was drawn by a chorus of alarm calls from the colony of Silver Gulls to a dark bird passing south over East Reef. Although closely pursued by the gulls it flew unhurriedly off to the west.

Australasian Gannet *Morus serrator*
Seen frequently offshore throughout the year, but most often between November and January, and generally out to sea within an arc from east to south-east or from west to south west. About half of the 151 records were of lone birds and the remainder were of groups of 2 to 30, usually flying in lines abreast or trailing. Immature gannets in mottled plumage were seen in all months, alone, in pairs and in the company of adults. Feeding dives were seen on a few occasions (see notes on Short-tailed Shearwater) and in May 1999, groups of 20 and 30 were seen feeding in conjunction with Crested Terns and Silver Gulls, on clupeoid fish probably being driven to the surface by Australian Salmon *Arripis truttaceous*.

Little Pied Cormorant *Microcarbo melanoleucos*
One bird, probably the same individual, was sighted on most visits through 1971 and 1972 (26 records) and was usually actively fishing. Lone birds were seen infrequently in 1969, 1973, 1976, 2001, 2002 and 2003 (12 records). Fishing activity was confined to the protected waters of the landing gutter, Seal Pool and the deep tide pools at the south end of Main Beach. Juvenile Bluethroat Wrasse *Notolabrus tetricus* up to 15

cm long were a common prey, and one other fish taken appeared to be a Crested Weedfish *Cristiceps australis*, which occurred in the permanent tide pools. When in the water the bird was extremely wary of seals and when resting always chose a rock well clear of those ashore. It became noticeably agitated if the resident Kelp Gulls passed overhead, but tolerated the close proximity of other resting cormorants.

Great Cormorant *Phalacrocorax carbo*
During the first decade of this survey this species was a rare visitor, in spring and summer, with only four sightings of one or two birds present for two or three days. During the afternoon of 24 November 1976, 6-10 birds settled near Black-faced Cormorants on North Plateau and towards evening a flock of 60+ flew past; the last and maximum count on North Plateau that season was 30 on 14 December 1976. Up to five were recorded at the roost in November and December 1977 and one to two in December 1978 and 1979. Single birds were recorded on North Plateau on visits in October and November 1999 and January 2000.

Little Black Cormorant *Phalacrocorax sulcirostris*
Four records only; of a single bird resting among a group of immature Pacific Gulls on Main Beach on 23 October 1972, of two birds resting near a Little Pied Cormorant and Black-faced Cormorants on Main Beach on 16 November 1972, of a single bird resting with Black-faced Cormorants on North Plateau on 27 December 2002, and a desiccated carcass on Main Beach below South Plateau on 6 January 1977.

Pied Cormorant *Phalacrocorax varius*
Six records in January, March, August and December of lone birds standing quietly on rocks close to the water's edge. One additional sighting was of two birds that perched briefly on the steel flying fox cable mid-way between the islets, on 7 March 1969. Not recorded in the second period of the survey.

Black-faced Cormorant *Phalacrocorax fuscescens*
In groups generally less than four, rarely more than 10, from February to November, roosting in the evening on portions of high rock masses not used by the fur seals, i.e. the south-east point of Black Rock and the north edge of North Plateau, with a distinct preference for the

latter. In December to January over seven seasons (1966-1972), numbers increased to an average of 26 and 32 respectively. A maximum of 67 was recorded on 14 December 1977. In the second survey period, numbers ranged from 0-57 and averaged 11.7.

On North Plateau they invariably roosted in a group along the northern edge, spaced uniformly about 1 m apart, standing or sleeping in an upright posture. Individuals sometimes tore up abandoned Silver Gull nests of dried noon flower stems, tossing beakfulls of the material into the air. Their roosting area became heavily fouled with white guano splashes during summer and encroached on nest sites of Crested Terns, Silver Gulls and a pair of Sooty Oystercatchers. Some birds were occasionally seen fishing in open waters near shore, but never in the Seal Pool or in the deep tide pools on Main Beach. Disgorged food items found in the roosting area on North Plateau included a small leatherjacket (Aluteridae) and an Australian Salmon.

Cattle Egret *Ardea ibis*
Single record of a solitary bird on Black Rock on 13 April 2000.

White-faced Heron *Egretta novaehollandiae*
Thirty-one records, mainly of lone birds, but also of groups of up to 10 in summer and autumn during periods of calm weather. About half the sightings were of birds flying past directly towards Phillip Island or north-west. Those seen on shore chose to roost either on the plateaux or outer reef areas well clear of any seals.

Royal Spoonbill *Platalea regia*
One sighting of a lone bird on 27 January 1966 that circled above Seal Rocks several times and then settled briefly on East Reef before heading north into a light north-easterly breeze.

White-bellied Sea-Eagle *Haliaeetus leucogaster*
One sighting on 16 November 1977 of a lone adult that made a low leisurely circuit of the islets followed by a mob of screaming Silver Gulls, that were at the peak of nesting. This bird was probably one of a pair resident at French Island and often seen in the vicinity of Sandy Point 17 km to the north (Davis and Reid 1975a).

Swamp Harrier *Circus approximans*
Four records of one or two birds passing — on 4 November 1969, 25 July 1971, 21 September 1971 and 18 January 1973. The resident Silver Gulls were greatly alarmed when they flew overhead and on one occasion a pair of nesting Kelp Gulls pursued a lone Harrier so closely that it was forced to flip and present its talons to break up the attack. The July record was of a lone adult flying slowly due south and steadily gaining in altitude until out of sight. A fifth record was of a lone bird perched on a wooden plank near the field station on 13 August 1969.

Nankeen Kestrel *Falco cenchroides*
Six sightings in January (1966, 1968 and 1972), and single sightings in April 1966, September 1968, November 1970 and June 1972. All were of lone birds either perched on cliff ledges or the railing of the observation tower's upper deck, or hovering over the plateaux. On 19 June 1972 a large female was disturbed from the carcass of a Common Starling from which most of the flesh had been stripped. She later returned and carried off the remains. Later that day two other partly consumed Starlings were found on Main Beach and South Plateau. The hind end of a small rat (possibly *Rattus norvegicus*), found on North Plateau two days later, had probably been carried to the islet by this Kestrel.

Brown Falcon *Falco berigora*
Four records; two birds on 20 January 1966, and lone birds on 26 February 1966, 17 January 1972 and 17 November 1977. On the first three occasions the birds remained high and showed interest in the activity of Silver Gulls and Crested Terns below, but they eventually flew off towards Phillip Island. On the last occasion the Falcon took a Silver Gull chick from West Beach and flew off to the north-east. Although the gull colony was greatly alarmed, the Falcon was not pursued.

Peregrine Falcon *Falco peregrinus*
Four records, possibly of the same bird, on 2 and 10 November 1970 and on 9 and 15 January 1971 circling the islets. On the latter two occasions it was driven from the vicinity of North Plateau by a mob of Silver Gulls and flew off to the Nobbies, rousing the gull colony there.

Australian Pied Oystercatcher *Haematopus longirostris*

Two birds were sighted on 4 and 16 November 1969 and lone birds on 7 October 1970 and 15 November 1972. All records were of birds making several circuits of the islets before heading off towards Phillip Island. On one occasion two birds settled on East Reef but were quickly driven off by a resident pair of Sooty Oystercatchers.

Sooty Oystercatcher *Haematopus fuliginosus*

Three resident pairs routinely nested on the islets throughout both survey periods. The maximum number recorded was nine on 25 October 1999 and there were less present in winter, usually four to six birds. The mean ($\pm$ s.e.) of 39 counts in the second survey period was 2.6 ($\pm$0.37). Until the arrival of a pair of adult Kelp Gulls in late 1970, two pairs nested on North Plateau and a third on Black Rock plateau. In subsequent seasons the Kelp Gulls ousted the latter, at a time when nesting by the other pairs was well advanced. The ousted pair then established a nest on the southern 'toe' of North Plateau, on two occasions appropriating a Silver Gull's nest. The North Plateau pairs were aggressively territorial, establishing their nests at least 35 m apart. During the entire span of 14 years these nest sites shifted by only a few metres. Nesting Silver Gulls were tolerated to within 3 m. Inter- and intraspecific interactions were common, and the latter were especially intense when parents were attending their recently hatched and highly mobile chick. On one occasion a sitting oystercatcher leapt up, caught and throttled a recently fledged Silver Gull that had ventured too near. The Gull was held down by a beak hold behind its head, the Oystercatcher standing motionless with its feet braced wide apart until the Gull was dead.

In 13 seasons the North Plateau pairs produced at least 25 clutches (14 × 2 eggs, 5 × 1 egg, 6 × eggs not visible). On four occasions two clutches were laid in a season, three after early failures, and in each case a chick was reared to fledging; in the fourth instance two eggs were laid after the fledging or loss of a near-fledged chick (c. 42 days after the first hatching), but the fate of the second clutch was not observed. In no case of clutches of two eggs were two chicks reared,

apparently because the first chick to emerge was moved by the parents and the second egg was abandoned in the nest — a sequence observed in two instances. Of the possible maximum production of 25 fledged young by the North Plateau pairs only 14 large 'runners' were actually found, partly because the parents were adept at hiding them in crevices and narrow spaces under boulders where their plumage blended perfectly with the black basalt. When we searched for these runners to band them, a parent would occasionally resort to a 'broken wing' display to lure the intruder away.

The success of the Black Rock pair was very difficult to follow, but in the five seasons prior to being ousted by the Kelp Gulls in 1972 they hatched at least three chicks. In the seven subsequent seasons when they nested on the southern 'toe' of North Plateau only five clutches were seen (3 × 2 eggs, 2 × 1 egg). In the summer of 1973/74 the Kelp Gulls abandoned Black Rock for Seal Rock and the Oystercatchers reclaimed their old nest site. They were first seen there on 10 January 1974 and four days later the nest contained one egg. The fate of this nesting was not observed. For the first four to five days after hatching chicks were fed small insects and arthropods probed for by their parents in the mats of noon flower, the chicks running from one to the other whenever something was captured. Thereafter chicks were fed with the flesh of chitons, small limpets, and univalve gastropods (*Nerita* sp.) garnered from the outer inter-tidal zones of Main Beach, East Reef and Middle Reef.

Adult Oystercatchers were extremely sensitive to the presence of humans and immediately vacated a nest if approached to within 50 m. This is similar to the flight initiation distance reported for this species by Glover *et al.* (2011). Despite great care to minimise disturbance, this was undoubtedly the cause of some nest failures as Silver Gulls were quick to plunder an egg and were suspected of taking at least one exposed newly hatched chick. It is likely that the poor success of the Black Rock pair when nesting on North Plateau was exacerbated by their close proximity to the field station located only c. 40 m away. One of these runners, banded on North Plateau 5 January 1977, was later found as a desiccated carcass on Forrest Caves beach, Phillip

Island on 24 April 1987. Another, banded on North Plateau 4 January 1980, was seen at Long Island Point near Hastings on 1 August 1981 and was subsequently trapped there on 12 June 1988 and released. Two resident colour-banded birds in the second survey period were banded respectively at Flinders (c. 10 km north-east) two years earlier and in Corner Inlet (c. 125 km south-east) six years earlier (as a two-year old).

Masked Lapwing *Vanellus miles*
A pair on Black Rock on 22 January 1968. One heard by KJC calling during night of 21-22 August 1969.

Ruddy Turnstone *Arenaria interpres*
In the first survey, Ruddy Turnstones were observed on all visits except during the winter of 1969. From March to August they were usually seen in groups of 5 to 10, their numbers then increasing substantially to about 50 in November to January. A few birds in richly coloured plumage were seen in April, July, August and September. In the second part of the survey, numbers ranged from 0 to 11 and averaged two birds per visit. Turnstones were observed on only 47% of visits in the second survey period.

Dispersed groups were often seen feeding along the shoreline at low tide, among weedy rocks and over beds of cunjevoi *Pyura* sp. Occasionally small flocks were seen moving about on, and the slopes above, the Main Beach and twice on South Plateau among clumps of noon flower where they appeared to be searching for insects. Towards evening and during strong winds they congregated to roost in sheltered parts of the shore platforms and often among the fur seals, where they moved about confidently and were totally ignored. Turnstones were often seen resting in very close proximity to Pacific Gulls.

Arctic Jaeger *Stercorarius parasiticus*
Thirteen sightings were logged of one to four individuals active offshore during spring and summer — from October to March in the first survey period. Most often they were noticed in November to January, when pursuing Silver Gulls returning from Phillip Island or the Mornington Peninsula with food for their young.

Fairy Tern *Sterna nereis*
Two or three birds seen offshore on three occasions by KJC in late December 1965 and a lone bird on 17 January1966.

Caspian Tern *Hydroprogne caspia*
Recorded only once, by KJC on 31 November 1968 feeding offshore. The Tern subsequently landed on North Plateau, but flew off when alerted by alarm calls of Silver Gulls.

White-fronted Tern *Sterna striata*
Commonly seen from late April to mid-November feeding offshore and often diving for small fish close in to the rocks, in the Seal Pool and in the shallow landing gutter. Often roosted overnight on Seal Rock, generally in groups of less than 10, but occasionally there were 50. They intermingled freely with roosting Crested Terns. On 16 June 1969 White-fronted Terns were seen diving into a swarm of krill to the south of Black Rock, where Fairy Prions, and Short-tailed and Fluttering Shearwaters were also feeding. Specimen RW#415 (Museum Victoria), stomach contained Coastal Krill; subcutaneous fat light orange in colour. This species was not seen on or in the vicinity of Seal Rocks during the second survey period but has been recorded occasionally along the southern coast of Phillip Island (Norman 1992) but not in Western Port between 1991 and 1994 (Dann *et al.* 2003).

Crested Tern *Thalasseus bergii*
Before the field station was erected on the South Plateau in December 1965, 40 nests were found on a narrow band of noon flower growing at the cliff edge, and at least eight pairs had nested on the north-west corner of North Plateau. After 1965 all nesting activity was concentrated on North Plateau, but only 20 nests were established. The output of this colony was about 12 young/yr until 1971/72 when the entire nesting was lost due to unusually heavy seal traffic on that plateau. In 1972/73 only three nests were found there, but eight nests were established on the upper boulder slope of West Beach among nesting Silver Gulls. During the next four seasons (1974/75 to 1977/78) nesting was confined to West Beach, but the maximum number did not exceed six. In 1974/75 and 1977/78 two

nests were established on the north end of South Plateau, which had not been visited during the period of courtship and nest-making in October, but occupation of the field station led to their failure. The last breeding recorded at Seal Rocks was of six nests in 1978 (Harris and Bode 1981). In 1994, a large colony became established at The Nobbies, 2 km north-west, and numbered 2050 nests by 2001 (Minton *et al.* 2001, Chiaradia *et al.* 2002).

Pairs engaging in high spiralling courtship flights were noted as early as 9 August and as late as 15 January. Some pairs slowly circled to heights of well over 300 m, came together and then plunged, one above the other, in a spectacular, slow-spiralling power-dive to within 10 to 20 m of the sea. Copulations were observed as early as 7 October and nests with eggs as late as 17 January. Chicks at the runner stage were moved to the shore platforms of North Beach and West Beach where they were protected by groups of adults, as many as three adults closely attending a single chick. Parents flying in with a fish were occasionally harried by Jaegers and on one occasion an adult with a bulging crop was pursued by a Kelp Gull. Fish up to c. 8-10 cm in length were fed to large chicks.

Crested Terns roosted throughout the year on both islets, but mainly on Seal Rock, usually as a single aggregation and often with White-fronted Terns. On Seal Rock their numbers varied erratically from less than 50 to more than 500, and in general the largest flocks occurred in late spring and summer. Exceptional concentrations were noted in 1972, of 1200–1500 on 16 November and of c. 2000 on 17 December. Roosting terns preferred the boulder and cobblestone-pebble areas of Seal Rock and the broad ledges of Black Rock, their selection depending on the strength and direction of the prevailing wind. During south-westerly gales terns congregated to roost on Main Beach in the lee of the South Plateau, where they crouched low and adjusted their orientation to any shift in wind direction.

Dead, injured and moribund adult Terns were found on Seal Rock from time to time and, on one occasion, an injured bird was attacked by Silver Gulls.

Pacific Gull *Larus pacificus*

These birds do not nest on the islets and were normally were seen resting on North Plateau, Main Beach and outer rocks, especially on East Reef. At least one pair of adults was noted on most visits in the first survey period, commonly there were three to five but not more than six. Counts of immatures varied more, from 2 to 50, with a tendency for larger numbers in winter. The maximum count of 50 on 9 February 1979 was of equal numbers of juveniles and immatures. This seasonal variation correlates with counts made in the vicinity of Sandy Point to the north (Davis and Reid 1975a). By contrast, Pacific Gulls (adults and immatures combined) were seen on only 60% of visits in the second survey period and never more than six were seen.

In strong winds Pacific Gulls sought protected areas, in particular the lower shingle slopes of Main Beach. They often competed with Silver Gulls for food items vomited by seals, and for fresh placentae during the pupping season, but were more watchful and tentative when in a close press of seals. Adult birds were seen to feed on the gastropod *Subninella* sp., which was common at the outer edge of the reefs, by dropping the shells from a height of 10 m or so onto the rocks to break them open.

A juvenile in dark plumage and with a badly injured wing was seen on Seal Rock on 14 November 1977. It was still alive on 28 March 1979, in sub-adult plumage, having survived by scavenging among the seals and tide-washed flotsam.

Kelp Gull *Larus dominicanus*

Lone adults were recorded on 22 August 1968 and 6 March 1969, and an adult pair arrived on 18 December 1970. They nested the following month on Black Rock and in all subsequent summers to 1979, when two additional pairs established nests. Other pairs were sighted on 29 October 1972 and 12 December 1977 but did not remain.

For some days after their arrival the founding pair was harassed by Pacific and Silver Gulls and by one pair of Sooty Oystercatchers, but this aggression quickly waned. Over the following eight seasons this pair produced a minimum of

14 eggs in eight clutches (1 × 3 eggs, 4 × 2 eggs and 3 × eggs not seen) from which 12 chicks were hatched and seven fledged. The progress of the 1971/72 nesting on Black Rock was not observed but a juvenile was seen in flight on 26 February 1972.

In the 1974/75 and 1978/79 seasons the original pair lost clutches when the nests were trampled by seals, and although a second nest mound of noon flower stems was raised nearby no eggs were laid. In 1973/74 and 1975/76 they abandoned their first nest (laying was not verified) and shifted to the other islet. In 1971/72 and 1977/78 a second nest was raised after the first brood fledged and in the latter season the first nest was renovated as well, but no eggs were laid. In January 1979 all three pairs laid, but two nests (clutch of two eggs in one; other's eggs not seen) were lost due to seal traffic. The third pair laid three eggs and two young were fledged. Laying dates varied from 28 October (estimated) to 13 January (observed), with most being laid between mid to late November.

Adult Kelp Gulls were dominant in any interaction with Silver and Pacific Gulls and routinely ousted a breeding pair of Sooty Oystercatchers from their favoured nest site on Black Rock. Physical clashes did not occur except for one instance on 15 November 1977 when a nesting adult killed a recently fledged Silver Gull

that ventured too near. Caught in a powerful beak hold behind the head, the young gull was jerked about violently and thrashed and then despatched with downward stabbing thrusts to the body. On 18 January 1973 the pair pursued and attacked a passing Swamp Harrier.

Kelp Gulls competed with other gulls for fragments of fish and squid vomited by seals, and for fresh placentae, and occasionally pecked at fresh seal carcasses. No predation of the eggs or chicks of other nesting birds was seen.

The mean number of Kelp Gulls frequenting Seal Rocks in the second survey period was 30.1 (+ s.e. 2.9) and ranged from 1 to 96 birds. There was some seasonal variation in numbers with generally higher numbers in spring and summer (Fig. 3). It was noticeable during winter that Kelp Gulls occurred in greater numbers across the southern shores of Phillip Island and on the Mornington Peninsula to the north (pers. obs.). In 2005 it was estimated that the number of breeding birds in 2002 was c. 50, and that pairs had started nesting on the western end of Phillip Island in 1995 (Dann 2007).

Silver Gull *Chroicocephalus novaehollandiae*

In the first survey period, this species was present all year and bred on the plateaux of both islets, on cliff ledges and amongst boulders on West Beach and Main Beach. The airspace beneath the floor of the field station was particularly favoured. Breeding pairs were estimated at 250 to 300, with no marked fluctuations from year to year. On 30 October 1978 Harris and Bode (1981) counted 192 occupied nests and 63 new but empty nests on Seal Rock. A marked post-breeding exodus occurred, with numbers dropping to as low as 20; however, the flock size between breeding seasons fluctuated erratically and occasionally exceeded 600. The mean number of Silver Gulls frequenting Seal Rocks in the second survey period was 177.3 (± s.e. 58.1) and ranged from 0 in August 2002 to 2000 birds in May 1999. Breeding was greatly reduced in the second survey period with fewer than 10 nests being found in most years usually above the field station built in 1979 and a few isolated nests on Black Rock. Silver Gulls have bred at the western end of Phillip Island since 1970s (Loyn 1975) and the colony has grown to about 2500 birds (PD pers. obs).

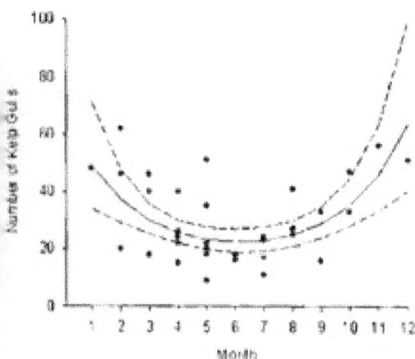

Fig. 3. Counts of Kelp Gulls recorded on Seal Rocks 1997 to 2005 by month. Numbers were greater in spring and summer. The quadratic function is significant (r = 0.8). The dashed lines are 95% confidence limits.

The commencement of egg-laying (on South Plateau) varied by as much as a month (27 July 1968, 21 August 1969, 21 July 1970, 27 August 1971) and the peak of breeding, in terms of the number of occupied nests, occurred 8-10 weeks later, at about the time the first fledglings were leaving the nesting area. Clutches were usually of two or three eggs, in nests formed of dried noon flower stems and a few feathers.

The fur seal colony provided a supplementary source of food during the latter part of the nesting season, when large numbers of pregnant females were ashore to give birth. Placentae and associated membranes from 2000+ births were shed between early November and mid-December, but only about half were thoroughly scavenged. Silver Gulls were quick to detect a female seal in labour or any seal showing signs of vomiting the remains of a previous meal. Gulls fed eagerly on any partially digested fish or squid and also on clotted masses of curdled milk that pups would occasionally vomit if trodden on by a large breeding male. Experience of this kind led some gulls to pounce on freshly voided seal faeces of an unusually pale cream colour, but they quickly stopped after a few beakfulls. Ever the opportunists, a group of gulls was observed avidly to devour a large mass of c. 80 mature tapeworms discarded after the dissection of a seal's digestive tract – behaviour that probably explained why strands of ripe proglottids were not found in any masses of soft faeces voided on shore. Individual gulls were attracted to and pecked at fresh wounds on fur seals and eagerly fed on clotted blood.

Insects were taken opportunistically. Gulls often snapped at blowflies *Calliphora* sp. attracted to seal carcasses and were adroit at catching moths *Agrotis* sp. which appeared in large numbers at various times on 17 October 1968, 4 November 1969 and 19 April 1977. They pounced on moths amongst the noon flower and pursued them when they swarmed into the air as high as 100 m. Larvae of the common cockchafer *Adoryphorus couloni* were fed to chicks by parents that followed the spring ploughing on Phillip Island and Mornington Peninsula. These grubs appeared to be an important source of food in some years. Undefended gull eggs were soon broken and devoured by other gulls, as were those of Crested Terns and Sooty Oystercatchers. Very small gull chicks that were displaced from the nest and not defended were pounced on by other adults, thrashed vigorously and then swallowed; unprotected larger chicks or 'runners' in down were pursued, buffeted and pecked about the head until they died.

Inadvertent disturbance by humans and by the fur seals contributed to this mortality and probably reinforced the tendency of parents to move their advanced chicks to the periphery of the nesting areas.

Silver Gulls were alert to any activity at sea nearby, flocking to investigate feeding by Crested and White-fronted Terns or the commotion caused by a Great White Shark *Carcharodon carcharias* preying on a seal at the surface, or a seal thrashing a large fish. On several occasions flocks of more than 100 gulls were observed feeding on surface shoals of small fish.

Apart from cannibalism only two instances of predation of Silver Gulls were observed— of an adult gull by an Eastern Barn Owl *Tyto javanica* on 5 October 1970, and of a chick by a Brown Falcon on 17 November 1977. On two occasions fledglings were killed when they ventured too close to nests of other species— by a Kelp Gull on 15 November 1977, and by a Sooty Oystercatcher on 25 November 1978. Silver Gulls were observed to harass and mob predators such as Eastern Barn Owl, Brown and Peregrine Falcons, White-bellied Sea-Eagle and Frigatebird, but were tolerant of Kelp and Pacific Gulls unless they approached occupied nests or runners.

Blue-winged Parrot *Neophema chrysostoma*
Two records of lone birds, on 19 March 1968 and 3 November 1969. The dates of these visits accord with the seasonality of sightings at Sandy Point (Davis and Reid 1975b). Specimen RW #454, (Museum Victoria) collected 3 November 1969, female, heavy mesenteric and subcutaneous fat deposits over abdomen and base of neck; ovary 8 x 5 mm, largest follicle 1.3 mm, crop empty.

Pallid Cuckoo *Cacomantis pallidus*
Seen only once, on North Plateau on 18 March 1968.

Research Reports

Eastern Barn Owl *Tyto javanica*
Six records and some prey remains, possibly representing the activities of three individuals.
A lone bird flew from the broken, stepped western edge of North Plateau on 10 June 1966 and settled in a crevice on the north face of Black Rock plateau. Regurgitated pellets were found below the south end of South Plateau the next day. This bird may have remained until at least 13 July, when the fresh remains of a Common Starling were found, picked clean in precisely the same manner as were Common Diving-Petrels taken by Barn Owls at Lady Julia Percy Island (RMW pers. obs.).
On 20 and 22 August 1970 a Barn Owl was seen successively at the rear window of the field station, roosting on Black Rock, and on a tank stand beside the station. On 15 September a Barn Owl flew from a recess below the northern point of South Plateau to East Reef, pursued by Silver Gulls; later it was on Main Beach directly beneath a steel cable that guyed the observation tower, incapacitated by a broken wing. On 5 October the fresh remains of a Silver Gull, headless and stripped of flesh, were found on West Beach, indicating another owl was in residence. This was confirmed two days later when a large Barn Owl flew from a roost below the South Plateau observation hide to Black Rock, pursued by Silver Gulls.

Sacred Kingfisher *Todiramphus sanctus*
One sighting by FTB, of a lone bird on Seal Rock on 10 November 1970.

Yellow-rumped Thornbill *Acanthiza chrysorrhoa*
One record on 19 November 1967 of a lone bird hopping about mats of noon flower on South Plateau; last seen flying to Black Rock.

Yellow-faced Honeyeater *Lichenostomus chrysops*
Lone birds were recorded on 21 December 1968 and 28 October 1975. The first sighting was of the bird battling against a strong wind near the field station and the second bird was seen perched on the roof of the South Plateau observation hide. Specimen: RW#361, 21 December 1968 (Museum Victoria), female, ovary c. 5 mm in length, largest follicle 1 mm; stomach contained four tiny flower(?) buds c. 2 mm in length.

White-eared Honeyeater *Lichenostomus leucotis*
One record of a lone bird flying about the eastern cliff face of South Plateau on 16 January1968.

Yellow-tufted Honeyeater *Lichenostomus melanops*
One record of a lone bird active about mats of noon flower near the rear of the field station on 15 May 1968. Closely observed with binoculars from a distance of about 6 m by KJC, FTB, WMB and RMW.

White-plumed Honeyeater *Lichenostomus penicillatus*
One record of a lone bird perched on the flying fox tripod, South Plateau, on 18 May 1971.

White-fronted Chat *Epthianura albifrons*
One record on 18 October 2000 on South Plateau.

Black-faced Cuckoo-shrike *Coracina novaehollandiae*
Single sighting of a lone bird on 19 April 1972, perched on the tower guys for about an hour.

Grey Shrike-thrush *Colluricincla harmonica*
Sighted once only on South Plateau on 27 April 1966.

Dusky Woodswallow *Artamus cyanopterus*
One sighting on 26 April 1966 by WMB of a lone bird perched on a ladder beside the door of the field station at dusk. It was captured after dark, found to be fit and was released next morning.

Raven *Corvus* sp.
One record of a flock of 12 birds flying past about 50 m offshore on 26 February 1971 and heading in a southerly direction. As none vocalised, identification to species was not possible. Presumed to be the locally common Little Raven *C. mellori*.

Satin Flycatcher *Myiagra cyanoleuca*
Recorded in most years in the first survey period during calm weather. The 27 records fall into two distinct seasonal groups: late February to March and November to mid-December, with a single record on 17 May of a female on West Beach. Pairs were seen twice in November, twice in December and once in March; the 23 other sightings were of lone birds. About half the sightings were of males. These flycatchers were always observed on or near noon flower on the edges and cliff faces of the South Pla-

teau, where they actively pursued insects. This species bred at Sandy Point (14 km north-east) in December 1962, 1963 and 1964 (Davies and Reid 1975c).

Magpie-lark *Grallina cyanoleuca*
One record on 23 November 1965 of a lone bird on Black Rock.

Rufous Fantail *Rhipidura rufifrons*
Seen once by WMB, on 13 December 1970, perched on the steps of the field station. WMB noted its white throat and broadly fanned russet tail. Davis and Reid (1975c) noted this species to be a rare visitor to Sandy Point.

Grey Fantail *Rhipidura albiscapa*
All the 40 records came from the first survey period and fell into two clear seasonal groups —autumn visits (mid-March to May) of groups of up to eight birds, and spring visits (mid-August to mid-December) of individuals and, rarely, pairs. Arrivals were always associated with calm weather and the fantails' activities were restricted to mats of noon flower on the plateaux and cliff faces, where they searched for flies and small moths. Usually, they left before the weather changed for the worse.

Willie Wagtail *Rhipidura leocophrys*
Two records of lone birds, on 17 August 1969 (KJC) active on South Plateau, and from 26 to 30 May 1976 when one was 'marooned' on the islet by very strong north-westerly winds. It confined its activities to the lee side of South Plateau, mainly about the base of the cliff.

Flame Robin *Petroica phoenicea*
Eighteen sightings during autumn on seven occasions during the first survey period—in mid to late March in 1966, 1968, 1969, 1972, 1973 and 1977, and on 13 and 14 May 1969. Generally, up to seven birds were seen at any one time over a period of one to seven days, and about three in four were in drab plumage. A dramatic influx to Seal Rock occurred on 19 August 1969 when at least 18 birds were in view (11 males, 7 in drab plumage). This was apparently part of the pre-breeding migration (see Davis and Reid 1975c). Specimens RW#434 female and #435 male, collected 20 April 1966, both very fat, testis of male 1.5 mm in length (Museum Victoria).

Australian Reed-Warbler *Acrocephalus australis*
Three records of lone birds active about the cliff faces of South Plateau, on 9 January 1969, 24 November 1969 and 17 June 1970. Specimens—RW #365, 9 January 1969 and RW #48, 17 January 1970 (Museum Victoria). The latter had bright yellow subcutaneous fat associated with the major feather tracts, and cream-coloured visceral fat; testis 3.3 mm in length.

Brown Songlark *Cincloramphus cruralis*
Recorded twice on South Plateau in mid-summer, on 14 January 1969 and 17 January 1970. The first bird was feeding on small butterflies, Australian Painted Lady *Vanessa kershawi* that had appeared in considerable numbers on the previous day, apparently from Phillip Island via a steady easterly breeze. Specimen: RW#369 (Museum Victoria), 14 January 1969, testis 1.5 mm in length; stomach contents remains of *V. kershawi*, an orange ichneumon wasp *Netelia* sp. and small flies.

Silvereye *Zosterops lateralis*
Nine sightings in the first survey period between October and May of one or two birds active about clumps of noon flower on the Seal Rocks plateaux. On two occasions, 2–4 November 1969 and 12–13 April 1973, repeated sightings may have been of the same individual or stragglers of larger groups passing through on pre- and post-breeding movements (see Davis and Reid 1975c).

Welcome Swallow *Hirundo neoxena*
A resident species, nesting in cavities and caves in the cliffs of South Plateau and Black Rock. One to two pairs seen throughout the year, usually skimming low over the upper beach slopes and plateaux. Very young nestlings were found on 2 November 1967 and 2 November 1969, and had fledged by the end of that month. Flocks of swallows were seen in December (1 × 9), January (1 × 30+), March (3 × 12+, 8, 10) and May (1 × 6).

Common Blackbird *Turdus merula*
Lone birds were sighted on Seal Rock on 21 April 1966 and 30 March 1979.

Common Starling *Sturnus vulgaris*
Small numbers bred regularly on Seal Rock,

pairs being noted as early as mid-August and nestlings being fed as late as 20 December, suggesting that several clutches were produced in a season. Only eight nests containing three to five eggs were found in well-hidden natural sites—in rock piles, in deep crevices in the cliffs, and one was behind a festoon of Ruby Saltbush on the cliff near the South Plateau observation hide; elsewhere nests were constructed in the roof ventilation shafts of the field station and in a corner of the South Plateau observation hide.

Parents actively foraged in clumps of noon flower and obtained an abundance of blowflies and their maggots from decomposing seal carcasses. Starlings were very wary of Silver Gulls, especially when gathering food in their vicinity, but on one occasion a parent defended a chick exposed to a threatening gull by landing on its back and pecking at its head.

Large flocks roosted on the cliffs overnight throughout the year, usually arriving at dusk from the direction of Phillip Island in small groups and separate compact flocks of 50-100 birds; occasionally a massed flock of 500+ birds was seen. Numbers appeared relatively similar between the two survey periods. As they approached these large flocks broke up into smaller groups which flew down at surprising speed straight onto the cliff faces. Incoming starlings were not deterred by strong winds and were often seen labouring against gale-force south-westerlies. A succession of small groups departed at or soon after dawn. Predation by a Barn Owl and a Nankeen Kestrel were noted.

Common Myna *Sturnus tristis*
A lone bird was seen on South Plateau on 25 November 1976 and a pair next day in the same vicinity.

House Sparrow *Passer domesticus*
A group of about five sparrows was seen on South Plateau near the field station on 10 June 1966, and lone birds were seen at the islet on 16 November 1966 and 16 November 1967.

Australasian Pipit *Anthus novaeseelandiae*
Six records of lone birds between mid-October and mid-April, active about the Seal Rocks plateaux and on Main Beach.

European Goldfinch *Carduelis carduelis*
Five records of 1-4 birds flying over or past

the islets, on 19 and 26 April 1966, 16 October 1968, 12 November 1968 and 26 July 1969.

Discussion

Although Seal Rocks lies a mere 1.5 km from Phillip Island, in character it is truly an island of Bass Strait. It is exposed to the full force of any gales and to the prevailing ocean swells from the south-west, which rise and break heavily around the two islets, and on occasion completely sweep the outer reefs and lower shore platforms. Powerful currents associated with the tidal flushing of Western Port contribute to turbulence of the sea to the north and east of Seal Rock. All the common Bass Strait seabirds adapted to these conditions were frequently seen in the waters adjacent to Phillip Island.

Seventy-five native and six exotic species of birds, including 21 native passerines, were observed at Seal Rocks or nearby. Six species were recorded breeding—Crested Tern, Silver Gull, Sooty Oystercatcher, Welcome Swallow, Common Starling and, for the first time in Victoria, Kelp Gull.

The variety of land birds recorded on Seal Rocks in the first survey period was surprisingly high, given the exposed nature and relative sterility of the terrain; however, the strait between Seal Rock and Phillip Island is narrow, all the species recorded there are common in the region, and most are wide-ranging seasonally or undertake significant north-south migrations. In each case the dates of sightings at Seal Rocks correspond with the timing of seasonal movements by the species concerned. Of particular interest is the clear evidence of departure and arrival of several small passerines known to migrate across Bass Strait—Satin Flycatcher, Grey Fantail and Flame Robin.

Changes over the 40-year period
There are a number of differences between the numbers and species of birds recorded in the first period of intensive observation (1966–1972) and the second period (1997–2005). Many fewer pelagic seabirds, Pacific Gulls and land birds were recorded in the second period and there were substantial changes in the numbers of breeding Crested Terns, Silver Gulls and Kelp Gulls. A comparison of the abundance and diversity of pelagic seabirds between the two periods is invalid as observations were not

made out to sea in the second period and hence relatively few seabirds were recorded. The preponderance of sightings of passerines and other land birds on South Plateau was no doubt a reflection of the activities centred on the field station there, which had the effect of deterring visits by seals, but may also have been related to the more extensive growth of noon flower on the plateau and its cliffs than elsewhere and the better shelter provided on its eastern face. With the removal of the field station on South Plateau in 1979, came much greater fur seal activity there and consequent destruction of noon flower areas on the top of the plateau. After 1997, vegetation was limited to a few sites inaccessible to seals on the sides of the plateau. This, together with the fact that the length of visits was substantially shorter in the second survey period meant that relatively few passerines were recorded there in later years.

The main changes to the breeding bird populations have been the movement and expansion of the breeding colonies of Crested Terns and Silver Gulls to adjacent parts of nearby Phillip Island and the arrival and expansion of the breeding Kelp Gull population. Crested Terns no longer breed on Seal Rocks; they have increased enormously in number locally and approximately 4000 birds breed on the Little Nobby (2 km east) (Chiaradia *et al.* 2002). The number of breeding Silver Gulls on Seal Rocks has decreased significantly over the past 40 years. Since the early 1980s there has been a consolidation of most of the Silver Gull colonies on the western end of Phillip Island to the two islets that make up 'The Nobbies' and the adjoining area of Point Grant. Disturbance from the increasing number of seals and Kelp Gulls may have been factors in both of these species moving from Seal Rocks and the subsequent increase may have been encouraged by the progressive elimination of foxes *Vulpes vulpes* in the western half of Phillip Island.

Kelp Gulls have also increased substantially at Seal Rocks since their arrival in 1968. Approximately 80 birds now breed there and they have successfully colonised a number of sites on nearby Phillip Island (Dann 2007) and on Lady Julia Percy Island (Dann *et al.* 2004). The expansion of Kelp Gulls at Seal Rocks may have

been associated with the expansion of fur seal numbers which, on one hand has reduced the number of suitable breeding sites, but on the other hand has increased the amount of food available in the forms of vomited food remains and placentae. In regard to the latter, the aggregate mass of placentae produced during the November–December pupping season would be at least 2000 kg and this represents an important source of high-quality protein for the gulls. However, many placentae are left unscavenged during the peak period of pupping, apparently because supply far exceeds demand. On the other hand, retching seals are always eagerly and competitively attended by gulls, and any ejected food items immediately snatched up by one or other of the three species of gull that occur on the islets.

Acknowledgements
The seal project was initiated and supported by generous grants from the MA Ingram Trust. KJ Chipperfield's interest in birds and his remarkably keen eyesight gave impetus to this study and stimulated the other team members to be alert and to record sightings. Collectively their observations formed the major part of this report. Janey Jackson painstakingly searched all the diaries and field journals to compile lengthy species dossiers summarised here. Victorian Wader Study Group provided details of banded oystercatchers seen and Roger Kirkwood organised the field trips in the second period, sharing his bird observations and commenting on a draft of this paper. Duncan Sutherland kindly prepared Figures 1 and 3 and assisted with the binomial generalised linear model.

Note
[1] Fred Baum (FTB), 1965–1979; Kevin Chipperfield (KJC), 1965–1970; Bill Bren (WMB), 1967–1973; Keith Cherry, 1970–1979; and Steve Craig, 1973–1977.

References
Chiaradia A, Dann P, Jessop, R and Collins P (2002) Diet of Crested Tern (*Sterna bergii*) chicks at Phillip Island, Victoria, Australia. *Emu* 102, 367–371.
Christidis L and Boles WE (2008) *Systematics and Taxonomy of Australian Birds.* (CSIRO Publishing: Collingwood, Vic).
Dann P, Arnould JPY, Jessop R and Healy M (2003) Distribution and abundance of seabirds in Western Port, Victoria. *Emu* 103, 307–313.
Dann P, Mackay M, Kirkwood R and Menkhorst P (2004) Notes on the birds of Lady Julia Percy Island in western Victoria. *The Victorian Naturalist* 121, 59–66.
Dann P (2007) The Population Status of the Kelp Gull *Larus dominicanus* in Victoria. *Corella* 31, 73–75.
Davis WA and Reid AJ (1975a) Western Port Report No. 1, Part 3. *The Victorian Naturalist* 92, 59–70.
Davis WA and Reid AJ (1975b) Western Port Report No. 1,

Part 4. *The Victorian Naturalist* 92, 124-123.

Davis WA and Reid, AJ (1975c) Western Port Report No. 1, Part 4 - continued. *The Victorian Naturalist* 92, 163-171.

Deagle BF, Kirkwood R and Jarman SN (2009) Analysis of Australian fur seal diet by pyrosequencing prey DNA in faeces. *Molecular Ecology* 18, 2022-2038.

Glover HK, Weston MA, Maguire GS, Miller KK and Christie BA (2011) Towards ecologically meaningful and socially acceptable buffers: Response distances of shorebirds in Victoria, Australia, to human disturbance. *Landscape and Urban Planning*, 103, 326-334.

Green RH (1973) Albatross Island. *Records of the Queen Victoria Museum* 51, 1-17.

Harris MP and Bode KG (1981) Populations of Little Penguins, Short-tailed Shearwaters and other seabirds on Phillip Island, Victoria, 1978. *Emu* 81, 20-28.

Kirkwood R, Gales R, Terauds A, Arnould JPY, Pemberton P, Shaughnessy PD, Mitchell AT and Gibbens J (2005) Pup production and population trends of the Australian fur seal. *Marine Mammal Science* 21, 260-282.

Loyn, RL (1975) Report on the avifauna of Westernport Bay. Project report; Westernport Bay Environmental Study, Melbourne; Ministry of Conservation, Victoria.

Minton C, Jessop R, Collins P and Graham D (2001) Tern breeding and banding 1999/2000 and 2000/2001. *Victorian Wader Study Group Bulletin* 24, 55-56.

Norman FI (1992) Counts of Little Penguins *Eudyptula minor* in Port Phillip Bay and off Southern Phillip Island, Victoria, 1986-1988. *Emu* 91, 287-301.

Page B, McKenzie J and Goldsworthy SD (2005) Dietary resource partitionsong among sympatric New Zealand and Australian fur seals. *Marine Ecology Progress Series* 293, 283-302.

R Development Core Team (2012) A language and environment for statistical computing, R Foundation for Statistical Computing, Vienna, Austria. URL: https://www.R-project.org/.

Shaughnessy PD (1978) Cape fur seals preying on seabirds. *Cormorant* 5, 31.

Warneke RM (1966) Seals of Westernport. *Victoria's Resources* 5, 44-46.

Warneke RM (1982) The distribution and abundance of seals in the Australasian Region, with summaries of biology and current research. In *Mammals in the Seas*, FAO Fisheries Series No. 5, Volume 4. pp. 431-475 (Food & Agriculture Organisation: Rome)

Warneke RM (2003) Seals at Seal Rocks, Western Port, and in Bass Strait, before and after the Baudin Expedition's visit in 1802. In *Le Naturaliste in Western Port 1802 2002 - Two Hundred Years of Change*. Proceedings of a seminar held at Cranbourne, Victoria, 13th and 14th April 2002, pp. 77-98. Ed N and P McWhirter, JL Sagliocco and J Southwood. (Department of Infrastructure/Mornington Peninsula Shire: Melbourne/Cranbourne)

Received 20 October 2011; accepted 13 September 2012

One Hundred Years Ago

Excursion to Phillip Island

BY JOSEPH GABRIEL

I am indebted to my co-leader, Dr. Brooke Nicholls, for the following notes on the bird-life of the outing. He says :— "As the result of several trips to Phillip Island just sixty species of birds have been recorded, but of these sixteen are sea or shore birds, leaving forty-four as residents of the island. These correspond very closely with the total of thirty-six species recorded in the *Naturalist* of December, 1911 (xxviii., p. 149), for the Bass Valley by Mr. A. W. Milligan and myself at Easter, 1911. The Bass Valley, it may be mentioned, is situated on the eastern side of Western Port, and at no great distance from Phillip Island. However, as each of our visits to the island and to the Bass Valley was made during the Easter holidays, observations at other periods of the year would doubtless add to the lists. The absence of the Spotted Ground-bird, *Cinclosoma punctatum*, Lath., from the Phillip Island list, and its inclusion in that of the Bass Valley, is perhaps the most interesting result of the comparison, and, while this bird has not yet been recorded for the islands of Bass Strait, it occurs in Tasmania. The presence of the Emu-Wren, *Stipiturus malachurus*, Shaw, the Orange-tipped Pardalote, *Pardalotus assimilis*, Ramsay, and the Mistletoe-bird, *Dicaeum hirundinaceum*, Shaw, upon the island is also of interest. Of the sea-birds found upon the island, the Short-tailed Petrel, or "Mutton-bird," *Puffinus brevicaudus*, Gld., and the Little Penguin, *Eudyptula minor*, Forst., bulk largest in importance. Both these birds are diminishing in numbers every year, and their rookeries are being gradually thinned out. It will be a surprise to many members of the Club to learn that the penguin is not upon the list of birds protected for some portion of the year. As Phillip Island is practically the last stronghold near the mainland of the Mutton-bird and the penguin, it is time they were afforded full protection in this locality. During the excursion some interesting observations were recorded regarding the penguins. The accompanying plate shows the nest of a pair of these birds, containing a young bird. The nest was some 500 yards inland from the sea, and placed high upon the cliff, amongst the tussocks. There were two openings to the burrow, which is unusual. In the foreground of the picture will be seen numbers of feathers scattered in front of the young bird. These are the shed feathers of the *second* down stage. During recent years it has been found that, many birds, especially penguins and petrels, shed two stages of down prior to acquiring the adult plumage. In the penguin the first down is of a fine, silky, hair-like structure. The young bird in the photograph had donned the adult plumage, which is attained prior to its leaving the nest and entering the sea."

From *The Victorian Naturalist* XXX, pp. 33-34, June 12, 1913

Sleeping aggregations of bees in relation to the risk of fire at their roosting sites in a forested, suburban landscape in eastern Australia

PJ Kubiak

PO Box 439, Ryde, NSW 1680, Australia

Abstract

Sleeping aggregations of at least 13 bee species (from the families Halictidae, Apidae, Colletidae and Megachilidae) were observed in the forested and fire-prone landscape of the Lane Cove River valley, in suburban northern Sydney, NSW, Australia, during the years 2002-2012. Bees were often found roosting at sites subjectively assessed as having a lower risk of being burnt. The fire risk of the observed sleeping aggregation sites may have been reduced by bees: 1. roosting in smaller vegetation patches, separated by a clearing from larger nearby areas of vegetation; or 2. roosting in areas of vegetation recently burnt by fire and therefore at a reduced risk of burning; or 3. roosting at or near the edges of vegetation, giving them a chance to escape into adjacent cleared areas, if a fire arrived when there was enough light for the bees to see and fly away; or 4. roosting at or near the edges of tracks or trails, which might act as fire breaks in the event of lower intensity fires; or 5. using combinations of some of the above four 'strategies'. This study suggests that sleeping aggregations of bees in this fire-prone area generally appeared to have a tendency to occupy roosting sites that were at a lower risk of being burnt, or sites that probably provided more opportunities for the bees to escape an approaching fire. There are a few indications in the published literature that some bee and wasp species in other fire-prone regions of the world may also have a tendency to occupy lower fire risk roosting sites. (*The Victorian Naturalist* 130 (1) 2013, 22 36).

Keywords: bee, fire, sleeping aggregation, communal roost, wasp

Introduction

Communal roosting has been observed in a number of insect groups (reviewed by Yackel Adams 1999), including butterflies (Lepidoptera) (Mallet 1986; Finkbeiner *et al.* 2012), bees and wasps (Hymenoptera) (see references below), dragonflies (Odonata) (Corbet 1999), beetles (Coleoptera) (Pearson and Anderson 1985; Webb 1994), flies (Diptera) (Allee 1927) and owlflies (Neuroptera) (Gomes-Filho 2000). Communal roosting has also been recorded for harvestmen (Opiliones) (Donaldson and Grether 2007).

The males (and occasionally females) of solitary bee species have been observed often gathering in the evening to sleep together at night, in both Australia and worldwide (Rau and Rau 1916; Rayment 1935; Linsley 1958; Evans and Linsley 1960; Linsley 1962; Michener 1974; Houston 1984; O'Toole and Raw 1991; Dollin *et al.* 2000; Alves-dos-Santos *et al.* 2009; and see images posted on the internet, e.g. www.australiannativebees.com). Similar behaviour has also often been observed in male and female wasps (Banks 1902; Bradley 1908; Rau and Rau 1916; Rau 1938; Evans and Linsley 1960; Linsley 1962; Evans and Gillaspy 1964; Callan 1984;

O'Neill 2001; Evans and O'Neill 2007). The term generally applied to communal roosting in bees and wasps is 'sleeping aggregation'. Male bees can form loose or dense sleeping aggregations, occasionally consisting of several species and ranging from a few bees to hundreds of individuals (Rayment 1935; Michener 1974; O'Toole and Raw 1991). Sometimes female bees may also be found sleeping near the males (Rayment 1935; Linsley 1962; Michener 1974). Typically, however, the females of most solitary bee species spend the night in nests, whereas the males of various species sleep together in communal roosts (Linsley 1958; Evans and Linsley 1960).

Bee sleeping aggregations tend to form towards the end of the day and, weather permitting, disband again the next morning (Evans and Linsley 1960; Linsley 1962; Alcock 1998). Roosting sites may be used by groups of male bees on successive nights for prolonged periods and the same sites are sometimes used by following generations of male bees in subsequent years (Evans and Linsley 1960; Linsley 1962; Alcock 1998; Wcislo 2003).

The most common type of sleeping aggregation probably involves male bees attaching themselves, either by the jaws and/or with their legs, to the stems or leaves of living or dead plants. Less commonly, males of various bee species may form sleeping aggregations in flowers, in communal burrows, under bark, in crevices or cracks, on seed pods and in bird nests (Rayment 1935; Linsley 1958; Cazier and Linsley 1963; Linsley and Cazier 1972; Raw 1976; Maynard 1991; O'Toole and Raw 1991; Azevedo and Faria 2007). In denser aggregations some of the bees may rest on top of each other, without contacting the substrate (Cazier and Linsley 1963). One of the intriguing aspects of these sleeping aggregations is that, whilst some male bees may compete aggressively with each other for mates during the daytime, the same individuals can be capable of peacefully roosting together at night (Raw 1976; O'Toole and Raw 1991).

The reason(s) for the formation of bee and wasp sleeping aggregations have apparently not been well understood (Rau and Rau 1916; Evans and Linsley 1960; Michener 1974; Dollin et al. 2000; Wcislo 2003; Alves-dos-Santos et al. 2009; Matthews and Matthews 2010). A number of researchers have put forward possible explanations for this phenomenon, often focusing on protection from predators and/or on thermoregulatory benefits (Rayment 1935; Rayment 1956; Evans and Linsley 1960; Linsley and Cazier 1972; Freeman and Johnston 1978; Callan 1984; Alcock 1998; Silva et al. 2011). A social function was suggested as a possible reason for sleeping aggregations in *Steniolia obliqua* wasps (Crabronidae) by Evans and Gillaspy (1964). However, it would appear that none of these explanations has been definitely proven (see Yackel Adams 1999, for a discussion of the possible function(s) and adaptive significance of communal roosting in bees and other insects). Similar and additional explanations have been suggested to account for aggregative behaviour in a wide range of animal species (Allee 1927; Ward and Zahavi 1973; Stephens and Sutherland 1999; Stephens et al. 1999; Marzluff et al. 1996; Bell et al. 2007; Grether and Donaldson 2007; Finkbeiner et al. 2012).

Matthews and Matthews (2010) considered that sleeping aggregations of bees are not in-

ternally organised and do not involve co-operative behaviour. However, it is possible that such aggregations might be an early step along the path towards the more co-operative behaviour of complex insect societies, as indicated by Rau and Rau (1916) and Rayment (1956). Aggregation pheromones may be involved in the formation of sleeping aggregations in bees and wasps (see Freeman and Johnston 1978; Alcock 1998; Wcislo 2003; Silva et al. 2011). Aggregation pheromones have been reported for a diverse range of non-social arthropods, including a few species of Hymenoptera (Wertheim et al. 2005).

Fire is important in shaping many terrestrial ecosystems in Australia and worldwide. Some researchers have studied the effects of fire on bee communities (Potts et al. 2003; Moretti et al. 2009; Grundel et al. 2010) and on individual bee species (Stow et al. 2007; Maynard and Rao 2010; Cane and Neff 2011).

The aim of this current study is to explore whether there might be a relationship between the roosting sites of bee sleeping aggregations and the risk of fire at those sites in the bushland of northern Sydney.

Study area

Observations for this study were made in the Lane Cove River valley of suburban northern Sydney, NSW, Australia. Surviving natural vegetation in the study area includes open-forest, tall forest, woodland, heathland, rainforest, riparian shrubland, mangrove forest, rushland and saltmarsh (Clarke and Benson 1987; Benson and Howell 1990; Benson and Howell 1994; Martyn 2010). Much of the surviving bushland in the Lane Cove River area is sclerophyllous, is situated on sandstone and has been broadly described by Keith (2004) as the Sydney Coastal Dry Sclerophyll Forests. This bushland has undergone varying degrees of fragmentation and the majority of the Lane Cove River catchment's natural vegetation has been cleared, for timber, agriculture and subsequently for suburban development, which has intensified in recent years. A more or less contiguous band of bushland survives along the course of the river and some of its tributaries. The largest areas of native vegetation occur in the upper reaches. Introduced weed species frequently dominate the study area's watercourses and also disturbed places, such as bushland edges. Even so, the study area still has a high diversity of native plant species.

Much of the vegetation in this study area could be described as 'fire-prone', in the sense that it is likely to be burnt quite frequently. The sclerophyllous vegetation is the most 'fire-prone', but areas containing rainforest species, mangroves, rushland and saltmarsh may also be burnt under extreme weather conditions. However, some patches of sclerophyllous bushland in the study area may escape being burnt for relatively long periods of time. Arson and planned fires set by bushland managers (for the purposes of hazard reduction and ecological management) are probably the two most common causes of bushfire in the Lane Cove River area in recent times. Occasionally, large wildfires have swept through the valley, e.g. in January 1994. Such fires can reach high intensities, depending on fuel levels in the bushland and weather conditions at the time of burning. Smaller bushfires occur fairly frequently in the Lane Cove River valley.

Methods

In the years 2002-2012 some bushland areas in the Lane Cove River valley were searched for sleeping aggregations of bees. Generally, searches were conducted in the late afternoon. The first aggregation was found by chance in 2002, when I was not looking for roosting bees. The pattern of searching tended to be biased towards looking along walking tracks, service trails and the edges of bushland because such places are easier to search. I attempted to counteract this bias by also searching bushland away from tracks and trails. Narrow tracks surrounded by thick, unburnt bushland were also searched and these were considered to be a very high fire risk situation for any bees that might have been found roosting along them. Several aggregations were found in a suburban garden in the vicinity of the Lane Cove River, located well away from the nearest bushland. A few bees (from eight species) were taken from a handful of the observed aggregations and sent for identification to Michael Batley, who also identified some bees from photographs. However, the bee species in the majority of the sleeping ag-

gregations were tentatively identified (without capturing the bees) by consulting Dollin *et al.* (2000) and by referring to the identifications, provided by Michael Batley, of similar looking bees. Some bees were not identified and these are grouped together as 'unidentified species' in Table 1. The number of bees in the smaller aggregations was counted, whilst bee numbers were estimated for larger aggregations.

The fire risk of each roosting site was subjectively assessed, taking into account characteristics such as the proximity and density of adjacent vegetation, the amount of leaf litter and other fuels present and the length of time since the last fire. Other factors that could potentially have modified the risk of fire to the bees were also noted, including whether the sleeping aggregation was situated on the edge of the bushland area, or next to a service trail or walking track. Roosting sites were given a subjective fire risk rating, ranging from very low to very high. Even when rain had recently fallen at a roosting site, the fire risk was assessed on the basis of what the risk would have been at the site under dry conditions. It was considered that, even in wetter periods, bushland could dry out quite quickly in the event of a run of successive hot, dry days without rainfall.

Results

Observations of bee sleeping aggregations made during this study are summarised in Table 1. At least 13 bee species (from the families Halictidae, Apidae, Colletidae and Megachilidae) were observed forming communal roosts. In some years more effort was put into searching for sleeping aggregations than in other years and this may largely account for variations in the numbers of aggregations found in different years. Generally, the number of bees found in roosts tended to peak in late spring to early or mid-summer. Sometimes sleeping aggregations persisted into late summer or autumn, but the numbers of bees aggregating at those times of the year were generally smaller. Of the small number of bees taken for identification, all were found to be males (M. Batley pers. comms.). Species from all of the bee families occurring in the study area were observed forming sleeping aggregations. Communal roosts of bees in the family Halictidae were the most frequently

Table 1. Some sleeping aggregations of bees observed in northern Sydney bushland, in relation to the subjectively assessed fire risk of their roosting sites. (Some roosting sites were in weed patches separate from or adjacent to natural vegetation). Aggregation size (maximum number of bees seen during the period of observation for each aggregation): v. small = 2–6 bees, small = 7–25 bees, medium = 26–50 bees, large = 51–200 bees, v. large = more than 200 bees (numbers were estimated for the larger aggregations), loose = a loose aggregation; Roosting substrate: (D) = bees roosted on a dead plant's skeleton, (L) = bees roosted on a living plant, (dt) = bees roosted on dead twig(s) of a living plant; Fire risk (modifying factors): E = roosted on edge of vegetation, te = roosted on edge of a track, RB = roosting site recently burnt; 'x'M = site near 'x' months prior to observation, B = roosting site burnt, sfb = roosted separate from bushland; 'x'm = roosted approx. 'x' metres from edge of bushland, garden = roosted in a garden (away from bushland); ? = uncertainty of species identification; * = an exotic or introduced plant species; (s.g.) = a member of the *Lipotriches flavovirulis* species group (probably *'L. excellens'*).

Family / Species	Date(s) of aggregation (month & year)	Aggregation size	Roosting substrate	Fire risk of roost site (subjectively assessed); risk modifying factors
Colletidae				
Euryglossa subsericea	Oct 2009	medium	*Melaleuca* sp. (L)	very low (garden)
?*Euryglossa subsericea*	Nov 2011	large	*Melaleuca* sp. (L–dt) (same site as Oct 09)	very low (garden)
?*Leioproctus thornleighensis*	Dec 2007	medium	*Kunzea ambigua* (L)	low to medium; te
Leioproctus thornleighensis	Jan 2011	medium	*Kunzea ambigua* (L) and *Grevillea sericea* (L)	high; E
Megachilidae				
Megachile ferox	Nov–Dec 2002	small	*Pimelea linifolia* (D)	low–RB?M; E
Megachile leucopygi	Nov 2010	small	*grass leaf (D) – weedy area	very low - sfb (c. 30m)
Megachile sp.	Oct–Nov 2007	very small	*weed (D)	low–sfb (c. 3m)
Apidae				
Amegilla ?bombiformis	Feb–Mar 2004	very small	*Kunzea ambigua* (D)	low - RB14M; te
Amegilla ?bombiformis	Feb–Mar 2005	very small	*weed (D)	low - sfb (c. 5 m)
Amegilla zonata (species group)	Mar 2003	small	*Juncus* sp. (L)	low - sfb (c. 5 m)
Amegilla sp. (banded bee)	Feb, Mar 2005	one bee	*weed (D) (with *A. ?bombiformis*)	low–sfb (c. 5 m)
Amegilla sp. (banded bee)	Jan 2006	very small	*Schoenus melanostachys* (L)	medium; te
Amegilla sp. (banded bee)	Jan 2007	very small	*weed (D)	low–sfb (c. 3–5 m)
Amegilla sp. (banded bee)	Jan 2008	very small	*Rumex* sp. (L)	low to medium; E
Amegilla sp. (banded bee)	Jan–Feb 2012	small	grass and *Euchiton* sp. (L)	low - sfb (c. 3m)
Thyreus ?nitidulus	Dec 05, Jan 06	very small (loose)	*Brassica* sp. (D)	low to medium; E
Thyreus ?nitidulus	Nov 2005	one bee	plant stem (D)	low to medium; E
Thyreus ?nitidulus	Jan 2006	one bee	*Allocasuarina littoralis* (D)	medium; E (above cleared area)
Thyreus ?nitidulus	Jan–Feb 2007	very small (loose)	*Brassica* sp. (D)	medium; E
Halictidae				
?*Homalictus punctatus*	Jan–Feb 2005	med–large	*Exocarpos cupressiformis* (L)	medium to high; te
?*Homalictus punctatus*	Dec 2005	very small	as above (same site)	as above
?*Homalictus punctatus*	Nov 2006	small	as above (same site)	as above

Table 1 (cont.)

Family / Species	Date(s) of aggregation (month & year)	Aggregation size	Roosting substrate	Fire risk of roost site (subjectively assessed); risk modifying factors
?*Homalictus punctatus*	Jan 2007	small	as above (same site)	as above
Homalictus punctatus	Oct–Nov 2007	small	as above (same site)	as above
Homalictus punctatus	Dec 2010	large	*Allocasuarina littoralis* (L)	high; E
?*Homalictus punctatus*	Feb 2012	very small	*Leptospermum trinervium* (L–dt)	low–B: c.12–24M; te
Lasioglossum subgenus *Australictus*				
?*Lasioglossum peraustrale*	Dec 2006	small	*Casuarina glauca* (L)	low
?*Lasioglossum peraustrale*	Dec 2006	medium	*Casuarina glauca* (L)	very low - sfb
Lasioglossum peraustrale	Dec 07–Jan 08	medium	*Casuarina glauca* (L)	low
?*Lasioglossum peraustrale*	Dec 10–Jan 11	small	*Acacia ?decurrens* (L)	medium; E
?*Lasioglossum peraustrale*	Dec 2011	very small	*Casuarina glauca* (L)	low
?*Lasioglossum peraustrale*	Dec 11–May 12	large	*Melaleuca linariifolia* (L)	medium; te
?*Lasioglossum peraustrale*	Jan–Mar 2012	medium	*Acacia mearnsii* (L)	high; E (above edge of clearing)
?*Lasioglossum peraustrale*	Jan–May 2012	small	*?Melaleuca armillaris* (L–dt) (planted)	very low–sfb (c.5 m)
?*Lasioglossum peraustrale*	Apr–May 2012	small	*Allocasuarina littoralis* (L)	low–sfb
Lasioglossum subgenus *Chilalictus*				
Lasioglossum convexum	Feb–Mar 2003	small	*Allocasuarina littoralis* (D)	very low - RB10M; tc
?*Lasioglossum instabilis*	Jan 2005	small	*Aristida vagans*	medium; te
Lasioglossum instabilis	Jan–May 2005	medium	*Themeda australis*	medium; near tc
Lasioglossum instabilis	Dec 05–Jan 06	large	*Bossiaea obcordata* (L) and *Themeda australis*	medium
?*Lasioglossum instabilis*	Jan 2006	very small	*?Digitaria* sp.	medium to high; te
?*Lasioglossum instabilis*	Jan–Mar 2006	medium	*Themeda australis*	medium; te
?*Lasioglossum instabilis*	Nov–Dec 2006	medium	*Themeda australis*	medium; near tc
?*Lasioglossum instabilis*	Dec 06–Jan 07	large	*Themeda australis*	medium; E
?*Lasioglossum instabilis*	Dec 06–Jan 07	large	*Themeda australis*	medium; E
?*Lasioglossum instabilis*	Dec 06–Jan 07	small	*Themeda australis*	medium; E
?*Lasioglossum instabilis*	Dec 06–Jan 07	small	*Themeda australis*	medium; E
?*Lasioglossum instabilis*	Dec 06–Jan 07	medium	*Dianella revoluta* (L–dt)	high; E
?*Lasioglossum instabilis*	Nov 06–Jan 07	large	*Themeda australis* (L) and *Aristida vagans*	medium to high; te
?*Lasioglossum instabilis*	Nov 2009	large	*Themeda australis* (L)	high; F
Lipotriches fortior	Jan–Feb 2011	large	*conifer (L–dt) - tree	low–sfb (c.4 m)
?*Lipotriches australica*	Nov–Dec 2002	small	*?Chlorophytum comosum* (L)	low – RB; E
Lipotriches australica	Feb 2003	very small	*Pomelea linifolia* (D)	very low - RB10M; F

Species	Date	Size	Plant	Notes
Lipotriches australica	Dec 2006	large	*Paspalum sp.	low-sfb
Lipotriches australica	Dec 2006	very large	*Cardiospermum sp.	low to medium; E
Lipotriches australica	Dec 06–Jan 07	large	*weeds (D) and *Verbena bonariensis (L)	low-sfb (c.10 m)
Lipotriches australica	Jan 2007	large	wire mesh fence - probably mowed from *Paspalum - see Dec 2006 (above)	very low
Lipotriches australica	Jan 2007	large	*Plantago lanceolata	medium; near E
Lipotriches australica	Dec 2007	small	*Paspalum sp. (L)	medium; F
Lipotriches australica	Jan–Feb 2011	large	*Verbena bonariensis (L)	low-sfb
Lipotriches australica	Jan–April 2012	very large	*Verbena bonariensis (L) & *Cardiospermum sp.	low to medium; E
Lipotriches australica	Jan–Feb 2012	large	*weed (D)	low to medium; E
?Lipotriches australica	Jan 2012	medium	*Verbena bonariensis (L)	low – sfb (c.10m)
Lipotriches australica	Feb 2012	medium	*Bidens pilosa (L)	medium; F
Lipotriches australica	Feb–Mar 2012	medium	*Sida rhombifolia (L) and *Paspalum sp.	medium -high; F
Lipotriches flavoviridis (s.g.)	Dec 2002	large	Pimelea linifolia (D)	very low – RB8M; F
Lipotriches flavoviridis (s.g.)	Jan–April 2003	small-med.	Pimelea linifolia (D)	very low – RB9M; T
Lipotriches flavoviridis (s.g.)	Feb–Mar 2003	small	Zieria smithii (D)	very low – RB10M; tc
Lipotriches flavoviridis (s.g.)	Feb–Mar 2003	small	Dillwynia retorta (D)	very low – RB2M; tc
Lipotriches flavoviridis (s.g.)	March 2003	very small	?Kunzea ambigua (D)	very low – RB3M; tc
Lipotriches flavoviridis (s.g.)	Jan 2005	large	*Plantago and grass (D)	medium; near F
Lipotriches flavoviridis (s.g.)	Jan 2005	medium	Pimelea linifolia (D)	medium to high; tc – above track
Lipotriches flavoviridis (s.g.)	Jan–Feb 2005	small	Themeda australis	medium to high; tc
Lipotriches flavoviridis (s.g.)	Feb–April 2005	medium	?Kunzea ambigua (D)	low – R26M; tc – above edge of track
Lipotriches flavoviridis (s.g.)	Mar 2005	very small	Juncus sp. (D)	low; F
Lipotriches flavoviridis (s.g.)	Sep 2005	one bee	Bursaria spinosa (L–dt) (site v. close to Feb 06)	medium; E (very close to Feb 2006 – see below)
Lipotriches flavoviridis (s.g.)	Nov–Dec 2005	large	plant (D) and Themeda	low; E – above edge of cleared area
Lipotriches flavoviridis (s.g.)	Jan 2006	medium	*Pavonia hastata (L) and Pimelea linifolia (L)	
Lipotriches flavoviridis (s.g.)	Nov–Dec 2005	very small	Pimelea linifolia (D)	medium to high; tc – above track
Lipotriches flavoviridis (s.g.)	Jan 2006	small	Themeda australis	medium; F
Lipotriches flavoviridis (s.g.)	Feb–March 2006	large	Bursaria spinosa (L) and Lepidosperma laterale	medium to high; tc – above track
Lipotriches flavoviridis (s.g.)	Nov–Dec 2006	small	Themeda australis	
Lipotriches flavoviridis (s.g.)	Jan 2007	small	Dianella revoluta (L–dt) (near ?L. instabilis)	high; F
Lipotriches flavoviridis (s.g.)	Jan–Feb 2007	small	*Bidens pilosa (D)	low to medium
Lipotriches flavoviridis (s.g.)	Jan–Feb 2007	large	*Sida rhombifolia (D) and *?Hypochaeris (D)	medium; near F
Lipotriches flavoviridis (s.g.)	Feb 2007	large	*Bidens pilosa (D)	high; E
Lipotriches flavoviridis (s.g.)	Nov 2007	medium	Dianella revoluta (L)	medium
Lipotriches flavoviridis (s.g.)	Nov 2007	medium	*Sida rhombifolia (L)	medium; E
Lipotriches flavoviridis (s.g.)	Nov–Dec 2007	medium	Acacia ulicifolia (L)	low – sfb
Lipotriches flavoviridis (s.g.)	Dec 2007	small	*Plantago lanceolata (L)	

Table 1 (cont.)

Family / Species	Date(s) of aggregation (month & year)	Aggregation size	Roosting substrate	Fire risk of roost site (subjectively assessed); risk modifying factors
Lipotriches flavoviridis (s.g.)	Dec 2007	small	*weed (D)	low - sfb (c.10 m)
Lipotriches flavoviridis (s.g.)	Dec 2007	large	*weed (D) and Senecio (L)	low
Lipotriches flavoviridis (s.g.)	Jan 2008	small	plant (D)	high; E
Lipotriches flavoviridis (s.g.)	Dec 2008	large	*weeds (D)	low
Lipotriches flavoviridis (s.g.)	Nov 2009	small	Dianella revoluta (L)	medium to high; E
Lipotriches flavoviridis (s.g.)	Dec 09-Jan10	large	*Sida rhombifolia	low
Lipotriches flavoviridis (s.g.)	Jan 2010	medium	*Bidens pilosa (D)	low to medium
Lipotriches flavoviridis (s.g.)	Jan 2011	large	*grass (L)	low to medium
Lipotriches flavoviridis (s.g.)	Mar 2011	small	*Coleonema pulchrum (L) - planted	very low (garden)
Lipotriches flavoviridis (s.g.)	Dec 2011	medium	*weed (D) (herb)	very low-sfb (c.4m)
Lipotriches flavoviridis (s.g.)	Dec 11-Jan 12	medium	Vimunaria juncea (L-dt)	high; E
Lipotriches flavoviridis (s.g.)	Dec 11-Jan12	large	*Sida rhombifolia (L-dt)	low to medium
Lipotriches flavoviridis (s.g.)	Dec 2011	small	Themeda australis	low
Lipotriches flavoviridis (s.g.)	Dec 11-Mar 12	large	Dianella revoluta (L-dt)	high; E
Lipotriches flavoviridis (s.g.)	Dec 2011	small	*Coleonema pulchrum (L) - same site as March 2011	very low (garden)
Lipotriches flavoviridis (s.g.)	Dec 11-Feb12	large	*Brassica sp (D)	medium; near E
Lipotriches flavoviridis (s.g.)	Jan-Feb 2012	very small	Themeda australis	high; near E
Lipotriches flavoviridis (s.g.)	Jan-Feb 2012	medium	Themeda australis (planted)	low-native garden-sfb
Lipotriches flavoviridis (s.g.)	Jan-Feb 2012	medium	Ozothamnus diosmifolius (L-dt)	low to medium; E
Lipotriches flavoviridis (s.g.)	Feb-Mar 2012	small	Lepidosperma laterale (L)	high; E
Lipotriches flavoviridis (s.g.)	Feb 2012	very small	Microlaena stipoides (L)	low-RB, E
Lipotriches flavoviridis (s.g.)	Feb-April 2012	small	Cymbopogon refractus (L)	high; E
Unidentified Species				
small, dark bee	Oct-Nov 2009	small	Ozothamnus diosmifolius (L)	high; L-above edge of cleared area
dark bee	Nov 2009	very small (loose)	Galvita sp. (L)	medium to high; tc
medium-sized, dark bee	March 2010	very small	*Coreopsis lanceolata (L) (flower)	low to medium
small, dark bee	Dec 2010	small	Austrostipa sp.	medium to high
small, dark bee	Dec 11-Jan 12	small	Allocasuarina littoralis (L)	low to medium; E
small, dark bee	Dec 11-Jan 12	medium	Austrostipa sp.	medium to high
small, dark bee	Jan-May 2012	large	Dillwynia retorta (L-dt)	high
small, dark bee	Jan 2012	one bee	*Hypochaeris sp (L) (flower) (roosting near L. flavoviridis s.g.)	
small, dark bee	Feb 2012	large	Dianella caerulea (L-dt)	low to medium; E
small, dark bee	Feb April 2012	large	Allocasuarina littoralis (L) - possibly planted	high
dark-coloured bee	Feb 2012	very small	*Hypochaeris sp.	low-sfb

observed and, within this family, bees of the *Lipotriches flavoviridis* species group (probably '*L. excellens*') were most often seen aggregating to roost together at night.

Bees were found roosting mostly on vegetation, including living grasses, sedges, herbs, shrubs and trees. Also, some bees roosted on the skeletons of dead plants or on dead twigs attached to living plants. One aggregation of *Lipotriches australica* spent some time roosting on a wire mesh fence. Bees roosted mostly at a height of less than two metres and often less than one metre above ground level. A few mixed aggregations, comprising several species, were observed, but most of the sleeping aggregations consisted of a single species. The majority of the observed sleeping aggregations were compact or dense, that is to say that most of the bees in each aggregation slept in close proximity to each other, often resting in contact with one or more of the other bees at the roost. Some of the sleeping aggregations apparently lasted for a longer time than others. However, some aggregations were observed for only one or a few days, so the actual duration of those aggregations was not determined. Also, the initial formation of an aggregation was rarely observed, so most of the aggregations probably existed for some time before they were first detected.

Bees often roosted in places that were apparently at a lower risk of being burnt than many other potential roosting sites in the surrounding landscape. Some of the bees formed sleeping aggregations in smaller patches of vegetation that were separated by a clearing from larger, nearby areas of denser bushland. For example, in December 2006 a sleeping aggregation of *Lipotriches australica* (Halictidae) (see front cover) was found on a relatively small patch of herbaceous weeds (Fig. 1) that was separated from the nearby larger area of bushland by a mowed, grassy clearing. The nearby open-forest had not been burnt for many years and no bees could be found roosting in this relatively high fire risk bushland. Another example occurred in December 2007 when a small aggregation of *Lipotriches flavoviridis* (species group) bees (Fig. 2) roosted on *Plantago lanceolata*, near the base of a remnant eucalypt (Fig. 3), separated from the nearby, dense bushland by a mowed

grassy area. Other similar examples included the aggregation of *Lipotriches fortior* found roosting on an exotic conifer in January/February 2011, separated from bushland by a cleared area and also several aggregations of ?*Lasioglossum peraustrale* (Halictidae) that roosted on trees separate from nearby bushland.

Bees sometimes roosted in areas of recently burnt vegetation located close to areas of unburnt bushland. For example, in November / December 2002 an aggregation of *Megachile ferox* (Megachilidae) bees roosted on the edge of an area of open-forest that had been burnt by a small fire approximately seven months earlier. The fire risk of this roosting site was relatively low because sufficient time had not elapsed since the recent fire for fuel loads to build up again. While *M. ferox* were roosting at this site, a wildfire burnt a larger area of the nearby adjoining forest, much of which had not been burnt for a long time. This fire burnt to within several hundred metres of where the bees were roosting. *Megachile ferox* continued to roost at the same site for several weeks after the occurrence of the nearby wildfire.

When bees roosted in places of higher fire risk they were often found at or near the edges of the bushland. This positioning may have given them the chance to escape into the adjacent cleared areas in the event of a fire, provided that such a bushfire occurred when there was sufficient light available and the bees were alert enough to fly away. Bees sometimes roosted on or near the edge of walking tracks. Some of these tracks could have acted as fire breaks in the event of lower intensity fires.

The apparent tendency to roost in lower fire risk locations may have varied somewhat between species. For example, some of the smaller bee species apparently tended to roost more frequently in places with a relatively higher risk of being burnt. However, more observations would be required to confirm whether this is generally the case. No bees were observed forming sleeping aggregations at sites considered to be at a very high risk of being burnt, even though narrow tracks that had thick bushland growing on either side of them were searched and some areas within denser vegetation were also searched.

Fig. 1. *Lipotriches australica* roosting site, December 2006 (bees are on band of weeds in the right foreground).

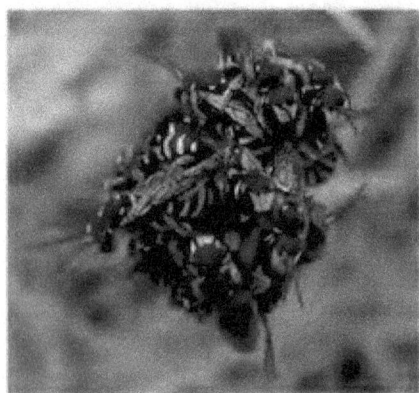

Fig. 2. *Lipotriches flavoviridis* (s.g.) sleeping aggregation, December 2007.

Discussion

These observations of sleeping aggregations over a period of a decade indicate that, in this fire-prone study area in south-eastern Australia, bees generally appear to occupy roosting sites with a lower risk of burning. Many other researchers have observed sleeping aggregations of bees and wasps. Some of these studies were conducted in fire-prone landscapes and provide a few indications that bees and wasps in other parts of the world may also tend to occupy lower fire risk roosting sites in fire-prone regions.

Bees and wasps roosting in sites possibly protected from fire

Bradley (1908) found a large concentration of sleeping aggregations, mostly of wasps, but including three bee species, in California, USA. His observations were made during summer in the San Joaquin Valley, when the vegetation had been parched by more than a month of hot, dry conditions. The large numbers of wasps (and a few bees) were aggregated at intervals along a road for 'perhaps a mile or more'. The roosting

Fig. 3. *Lipotriches flavoviridis* (s.g.) roosting site, December 2007 (bees are in bottom right hand part of photo).

groups of wasps and bees were scattered along a narrow strip of dried vegetation between the road and a recently harvested grain field. On the other side of the road from the sleeping aggregations of wasps and bees, the vegetation had been recently burnt by 'extensive prairie fires'. Bradley searched the sagebrush and foxtail grass on the plains 'twelve miles distant', but was unable to find any other sleeping aggregations of wasps. This suggests that these wasps and bees may have selectively occupied a roosting site that was at a lower risk of burning, compared with alternative sites on the plains.

In this present study sleeping aggregations of bees were sometimes found on smaller vegetation patches, separated by a clearing from larger, nearby areas of vegetation. This behaviour may have protected the bees from the greater risk of fire involved in roosting in the larger areas of vegetation, just as the wasps (and bees) that Bradley observed in the San Joaquin Valley may have received some protection by roosting in a narrow roadside strip of vegetation. Some of the bees observed in my study roosted at or near the edges of tracks or trails. This may possibly have given them some protection from fire, as some of these tracks could potentially have acted as fire breaks in the event of lower intensity fires. However, this would probably not apply to very narrow tracks surrounded by thick, fire-prone vegetation. Price and Bradstock (2010) found evidence to indicate that roads may stop some fires in the dry sclerophyll forests of the Sydney region.

Evans and Linsley (1960) and Linsley (1962) studied a diverse array of bee and wasp species

gathered together in a concentration of sleeping aggregations in the Chiricahua Mountains, in Arizona, USA. The site of these aggregations was a meadow, approximately 30 m × 90 m (ca. 100 × 300 feet), situated opposite a building, across an access road and car parking area, at a research station. The brief site description provided by the authors suggests that the meadow may have been somewhat protected from fire, at least on one side. This meadow roosting site occupied by these bees and wasps may have served as a refuge from fires, when compared with the woodlands surrounding the research station, which may have been more likely to be burnt than the meadow. The vegetation of the Chiricahua Mountains is prone to fires and there was recently a major wildfire in the area.

Bees roosting in recently burnt areas

Rau and Rau (1916) found two sleeping aggregations of male *Svastra obliqua* (as *Melissodes obliqua*) (Apidae) bees roosting on scorched leaves in recently burnt areas in open fields in Missouri, USA, but they were unable to find bees of this species roosting in nearby unburnt vegetation. Frankie *et al.* (1980) studied the bee *Centris adani* (Apidae) at a site, consisting of farmland interspersed with patches of regrowing dry forest, in Costa Rica. Their study site included areas of unburnt tall grass and 'brush', but the only sleeping aggregation of male bees that they found was located in a recently burnt area. In this present study, an aggregation of *Megachile ferox* (Megachilidae), some aggregations of *Lipotriches flavoviridis* (species group) (Halictidae) and one aggregation of *Amegilla*

?*bombiformis* (Apidae) bees were found roosting in recently burnt areas.

Before European settlement there were large areas of fire-prone contiguous vegetation in the Sydney district and roosting in recently burnt areas may have been an important way for bees to avoid the hazard of fire. Also, bees may have used vegetation associated with larger rock outcrops as roosting sites, because such situations may have offered some protection from fire. Even though much of the bushland closer to Sydney is now fragmented, it seems likely that recently burnt areas may still offer roosting bees a refuge from the risk of fire in those suburbs of the city that have retained some remnant natural vegetation (such as in the Lane Cove River valley).

Schowalter (2000) indicated that some insects with longer (2–5 year) generation times may avoid places where litter has accumulated in fire-prone ecosystems. Miller and Wagner (1984) found that pupae of the Pandora Moth *Coloradia pandora pandora* (Saturniidae), in a pine forest in Arizona, tended to occur in greater numbers where fuel loads were lighter on the ground and the canopy was more open. They speculated that this tendency may have reduced the risk of the pupae being killed by fire. Frost (1984) stated that some bird species apparently prefer to nest on recently burnt ground and suggested that these birds might be able to detect predators more easily and that predators might be less plentiful in recently burnt areas. The eggs, chicks and nesting adults of these birds would also be less exposed to the risk of their being killed by fire, as recently burnt areas would be less likely to support fires than unburnt areas. Frost noted that some of these bird species have dark coloured eggs and chicks and suggested that these might be 'adaptations' to nesting on blackened ground.

Rau and Rau (1916) observed that the brown bodies of the bee *Svastra obliqua* (Apidae) blended very well with the 'dingy' burned leaves of the scorched plants on which some of these bees roosted in the two recently burnt areas mentioned above. In this current study it was noted that the generally brownish colouration of *Lipotriches flavoviridis* (species group) (Halictidae) bees blended effectively with the brownish stems of the scorched *Pimelea lini-folia* and *Zieria smithii* skeletons that some of these bees roosted upon in recently burnt bushland in 2003. This 'camouflage' was particularly effective at lower light levels and when the bees were motionless (and in smaller numbers). This bee species also sometimes roosts on the brownish skeletons of dead plants in areas that have not been recently burnt, as well as on the brownish and green portions of some living plants, in the Lane Cove River valley.

Bees and wasps roosting at the edges of vegetation

Researchers have reported finding bees and wasps roosting at or near the edges of vegetation. For example, Rau and Rau (1916) noted that a thick 'mass' of weeds bordering a large open area seemed to be a favoured roosting site for the wasp *Myzinum* sp. (as *Elis 5-cincta*) (Tiphiidae) in the USA. Also, Rau (1938) observed sleeping aggregations of the wasp *Prionyx atratus* (as *Prionomyx atratum*) (Sphecidae) on weeds at the edge of a harvested wheat field and the wasp *Ammophila nigricans* (Sphecidae) roosting on low plants between a garden and 'the woods', in Missouri, USA. Mathewson and Daly (1955) found the bee *Melissodes denticulata* (as *M. perplexa*) (Apidae) aggregating to sleep on *Verbena stricta* in a 'weedy clearing' in Kansas, USA. More recently, Hausl-Hofstätter (2008) found a small number of individuals of 8 bee species and 3 wasp species, over a number of years, roosting in a forest clearing, next to a roadside in Croatia. Such observations are similar to those made in this present study, of bees forming sleeping aggregations at or near the edges of vegetation in the Lane Cove River valley. In this study area the remaining fragments of native vegetation have long perimeters, and potential roosting sites at the edges of vegetation are abundant. Observations during the course of this study of apparently suitable roosting sites within unburnt vegetation, away from the edges, indicated that they were mostly unused by bees.

There could be a number of reasons why bees and wasps might favour the edges of vegetation for their roosting sites. For example, they may prefer to roost in more open areas because such places could be more likely to be exposed to sunshine, enabling the insects to use the sun's heat to warm themselves before taking off in the morning and/or to keep them warm whilst settling in the afternoon. The potential significance of such thermoregulatory benefits could be illustrated by the observation that sleeping in dark flowers evidently enables some bee species to warm up more quickly in the morning (Dafni et al. 1981; Sapir et al. 2005; Sapir et al. 2006). Kaiser (1995) observed that several roosting solitary bee species became active only after they were exposed to direct sunlight. Another possible benefit of roosting at or near the edge of vegetation is that bees and wasps might be able to flee into the adjacent cleared area, if a fire happened when there was enough light for them to see and fly away. On very warm, sunny days, some bees observed in this study took a long time to settle down at their roosting sites, in the late afternoon or evening, before ceasing activity at or before sunset. On such days, there could be quite a long time when the settling bees would be alert enough to respond to any fire in the vicinity of their roost.

Oldroyd and Wongsiri (2006) noted that the location of drone congregation areas of Asian honey bees is apparently governed by particular physiographic features of the landscape that attract the males. Roosting solitary bee species may also follow simple rules to decide where to roost. For example, some bees and wasps may prefer more open roosting sites. Such a preference may have originated from a possible thermoregulatory benefit provided by roosting in locations more exposed to sunshine, as discussed above. The possible added benefit of a lowered risk of fire may have, in turn, reinforced the initial preference for roosting in more open areas.

The potential reaction of roosting bees to an approaching fire

It is not clear exactly how roosting bees would react to an approaching fire. Visscher et al. (1995) quoted reports that worker Honey Bees *Apis mellifera* (Apidae) respond to smoke by engorging themselves with honey drawn from the comb and that smoke reduces the number of guard bees at the hive entrance. They found that applying smoke to isolated *A. mellifera* antennae reduced the response of these antennae to honey bee alarm pheromones. Oldroyd and

Wongsiri (2006) noted that people in southeast Asia, who climb trees for wild bee honey at night, will strike the tree branch near the nest with a torch of bundled burning leaves and that some disoriented bees from the nest will follow the falling sparks, as they cascade downwards. They also considered that hunters using smoke to drive the wild Asian honeybee *Apis dorsata* away from nests during the daytime, before harvesting honey (rather than burning them off at night) are likely to greatly increase the chances of the bee colonies surviving. Frost (1984) noted that ticks may respond to smoke by dropping to the ground to seek shelter and that some grasshoppers are apparently capable of fleeing some fires. Schütz et al. (1999) indicated that the Colorado Potato Beetle *Leptinotarsa decemlineata* (Chrysomelidae) responds to high concentrations of fire-generated volatile compounds by exhibiting 'escape behaviour'. If bees can detect smoke from fires, then roosting bees might attempt to fly away from an approaching fire, if it occurred when there was sufficient light for the bees to see. If this is the case, then bees (and wasps) roosting next to a cleared area would probably have a better chance of surviving an intense, fast moving, daytime fire than if they had roosted at a site surrounded by thick vegetation.

Further ways roosting bees and wasps may reduce fire risk

Rau and Rau (1916) observed *Chalybion californicum* (as *C. caeruleum*) (Sphecidae) wasps aggregating under a rock overhang and Rau (1938) found the same species forming sleeping aggregations in an abandoned house, in the USA. Evans et al. (1986) noted that males and females of the sand wasp *Bembecinus quinquespinosus* (Crabronidae) slept in clusters under rocks in Colorado, USA. They suggested that the rocks may have protected the wasps from cooler night-time temperatures (and the wind) and helped the wasps to warm up in the morning. It is possible that sleeping under rocks may also provide some wasp species with a degree of protection from fire that they would not receive while roosting on vegetation. In this case, thermoregulation could be the primary benefit and protection from fire might be a secondary benefit of roosting under rocks.

Evans and O'Neill (2007) found that males of the sand wasp *Bembix cursitans* (Crabronidae) constructed unusually deep sleeping burrows at a site in coastal Western Australia. Such burrows could provide the males of this wasp species with protection from fire, at night.

Cane and Neff (2011) indicated that ground-nesting bees may tend to prefer bare patches of ground or soil banks for their nest sites and that this could provide some protection from the heating effects of wildfires, due to the lower levels of fuel at such sites. Other insect species may find more unusual refuges from fire in the landscape. For example, Brennan *et al.* (2011) observed that some insects and other invertebrates survived in experimentally burnt grass trees *Xanthorrhoea preissii* (Xanthorrhoeaceae) in Western Australia.

In fire-prone vegetation, such as much of Sydney's bushland, fire probably poses a significant risk to bee populations. During the daytime, bees might be able to evade fires, depending upon their ability to detect cues such as smoke early enough for them to fly away from the path of an approaching fire. The faster and more intense fires probably pose a greater risk to bees and other insects. The females of some solitary bee species probably receive protection from fires at night because they make their nests by burrowing deeply enough into the ground to avoid over-heating in the event of a fire (Potts *et al.* 2003; Cane and Neff 2011). However, the females and young of some other solitary bee species are probably at a greater risk of being killed by fires because their nesting locations (e.g. inside thin plant stems) are not adequately protected from high intensity fire (Potts *et al.* 2003; Maynard and Rao 2010; Cane and Neff 2011). The males of most bee species in the Sydney region probably sleep on vegetation, in the open, at night. There has probably been selection pressure on roosting bees in the Sydney region, over many centuries, favouring the survival of bees that tend to occupy roosting sites that are at a lower risk of being burnt, or sites affording more opportunity for the roosting bees to flee an approaching fire. This may also apply to bees and wasps in fire-prone environments in other parts of the world.

Further research possibilities

These observations over a ten year period indicated that roosting bees in the study area generally tended to occupy sites at the safer end of the fire risk spectrum. The bee roosting sites appeared not to be randomly distributed throughout the landscape and fire risk reduction appeared to be a likely factor in the location of bee roosting sites in this fire-prone landscape. However, it may be difficult to separate this effect from the possible thermoregulatory benefits gained by bees roosting in more open areas and on the edges of vegetation. Additional field work, possibly supplemented by experimental studies, could help to determine the relative influence of these factors on bee roosting site selection. Another avenue of research could be the response of roosting solitary bee species to approaching fire and the effects of smoke on bee sleeping aggregations.

Concluding remarks

This study indicates that solitary bee species observed forming sleeping aggregations in the fire-prone Lane Cove River valley, in south-eastern Australia, appeared to have a general tendency to roost in sites that were at a lower risk of being burnt, or sites likely to offer a greater chance for the bees to escape a threatening fire. Some bee and wasp species in other fire-prone regions of the world may also show a tendency to occupy lower fire risk roosting sites. The possibility that some roosting bees and wasps occupy sites at a lower risk of burning in fire-prone ecosystems is an idea providing opportunities for further research in both Australia and overseas. More field and experimental work could be required to unequivocally determine whether this idea is valid. There are also unanswered questions surrounding the wider phenomenon of the formation of sleeping aggregations in bees and wasps. The function of such sleeping aggregations is apparently not well understood. It is not clear whether there is one function or whether there are multiple reasons for this behaviour.

Acknowledgements
I'd like to thank David Robinson for organising permission from Ryde Council to collect bees in a council reserve. Thanks to Michael Batley for identifying eight bee species from specimens, for gener-

ously sharing his knowledge of these bees, advising on the taxonomic status of *Lipotriches flavoviridis* (species group) and *Amegilla zonata* (species group) and for identifying a number of bee species from photos. (However, I am responsible for any errors in bee identification or nomenclature that may be contained in this paper). Thanks also to David Britton for providing information about *Lasioglossum peraustrale* and *Euryglossa subsericea* and for communicating M. Batley's identifications of these species, and thanks to Emma Gray for conveying several of the earlier identifications to me. I'd also like to thank the two anonymous referees for providing interesting comments on my paper. (Note: The reference used to check nomenclature of sphecid wasps was 'Catalog of Sphecidae', compiled by W.J. Pulawski, www.research.calacademy.org/ent/catalog_sphecidae)

References

Alcock J (1998) Sleeping aggregations of the bee *Idionelissodes duplocincta* (Cockerell) (Hymenoptera: Anthophorini) and their possible function. *Journal of the Kansas Entomological Society* 71, 74–84.

Allee WC (1927) Animal aggregations. *The Quarterly Review of Biology*. 2, 367–398.

Alves-dos-Santos I, Gaglianone MC, Naxara SRC and Engel MS (2009) Male sleeping aggregations of solitary oil-collecting bees in Brazil (Centridini, Tapinotaspidini, and Tetrapediini; Hymenoptera: Apidae). *Genetics and Molecular Research* 8, 515–524.

Azevedo AA and Faria LRR (2007) Nests of *Phacellodomus rufifrons* (Wied, 1821) (Aves: Furnariidae) as sleeping shelter for a solitary bee species (Apidae: Centridini) in south eastern Brazil. *Lundiana* 8, 53–55.

Banks N (1902) Sleeping habits of certain Hymenoptera. *Journal of the New York Entomological Society* 10, 209–214.

Bell WJ, Roth LM and Nalepa CA (2007) *Cockroaches: Ecology, Behavior and Natural History*. (The Johns Hopkins University Press: Baltimore)

Benson D and Howell J (1990) *Taken for Granted: The Bushland of Sydney and its Suburbs*. (Kangaroo Press: Kenthurst, NSW)

Benson D and Howell J (1994) The natural vegetation of the Sydney 1:100 000 map sheet. *Cunninghamia* 3, 677–787.

Bradley JC (1908) A case of gregarious sleeping habits among aculeate Hymenoptera. *Annals of the Entomological Society of America* 1, 127–130.

Brennan KEC, Moir ML and Wittkuhn RS (2011) Fire refugia: the mechanism governing animal survivorship within a highly flammable plant. *Austral Ecology* 36, 131–141.

Callan EM (1984) Notes on a sleeping aggregation of *Priony x globosus* (F. Smith) (Hymenoptera: Sphecidae). *Australian Entomological Magazine* 11, 38.

Cane JH and Neff JL (2011) Predicted fates of ground-nesting bees in soil heated by wildfire: thermal tolerances of life stages and a survey of nesting depths. *Biological Conservation* 144, 2631–2636.

Cazier MA and Linsley EG (1963) Territorial behavior among males of *Protoxaea gloriosa* (Fox) (Hymenoptera: Andrenidae). *The Canadian Entomologist* 95, 547–556.

Clarke PJ and Benson DH (1987) *Vegetation Survey of the Lane Cove River State Recreation Area*. (Ecology Section, Royal Botanic Gardens: Sydney)

Corbet PS (1999) *Dragonflies: Behavior and Ecology of Odonata*. (Cornell University Press: New York)

Dafni A, Ivri Y and Brantjes NBM (1981) Pollination of *Serapias vomeracea* Briq. (Orchidaceae) by imitation of holes for sleeping solitary male bees (Hymenoptera). *Acta Botanica Neerlandica* 30, 69–73.

Dollin A (2001) Maureen's 'cluster of bees'. *Aussie Bee* 16, 4–6.

Dollin A, Batley M, Robinson M and Faulkner B (2000) *Native Bees of the Sydney Region: A Field Guide*. (Australian Native Bee Research Centre: North Richmond, NSW)

Donaldson ZR and Grether GF (2007) Tradition without social learning: scent-mark-based communal roost formation in a Neotropical harvestman (*Prionostemma* sp.). *Behavioral Ecology and Sociobiology* 61, 801–809.

Evans HE and Gillaspy JE (1964) Observations on the ethology of digger wasps of the genus *Stenifia* (Hymenoptera: Sphecidae: Bembicini). *The American Midland Naturalist* 72, 257–280.

Evans HE and Linsley EG (1960) Notes on a sleeping aggregation of solitary bees and wasps. *Bulletin of Southern California Academy of Sciences* 59, 30–37.

Evans HE and O'Neill KM (2007) *The Sand Wasps: Natural History and Behavior*. (Harvard University Press: Cambridge, Massachusetts)

Evans HE, O'Neill KM and O'Neill RP (1986) Nesting site changes and nocturnal clustering in the sand wasp *Bembecinus quinquespinosus* (Hymenoptera: Sphecidae). *Journal of the Kansas Entomological Society* 59, 280–286.

Finkbeiner SD, Briscoe AD and Reed RD (2012) The benefit of being a social butterfly: communal roosting deters predation. *Proceedings of the Royal Society B* 279, 2769–2776. doi: 10.1098/rspb.2012.0203.

Frankie GW, Vinson SB and Coville RE (1980) Territorial behavior of *Centris adani* and its reproductive function in the Costa Rican dry forest (Hymenoptera: Anthophoridae). *Journal of the Kansas Entomological Society* 53, 837–857.

Freeman BE and Johnston B (1978) Gregarious roosting in the sphecid wasp *Sceliphron assimile*. *Annals of the Entomological Society of America* 71, 435–441.

Frost PGH (1984) The responses and survival of organisms in fire-prone environments. In *Ecological Effects of Fire in South African Ecosystems*, pp. 273–309. Eds P de V Booysen and NM Tainton. (Springer-Verlag: Berlin)

Gomes-Filho A (2000) Aggregation behavior in the Neotropical owlfly *Cordulecerus alopecinus* (Neuroptera: Ascalaphidae). *Journal of the New York Entomological Society* 108, 304–313.

Grether GF and Donaldson ZR (2007) Communal roost site selection in a Neotropical harvestman: habitat limitation vs. tradition. *Ethology* 113, 290–300.

Grundel R, Jean RP, Frohnapple KJ, Glowacki GA, Scott PE and Pavlovic NB (2010) Floral and nesting resources, habitat structure, and fire influence bee distribution across an open-forest gradient. *Ecological Applications* 20, 1678–1692.

Hausl-Holstätter U (2008) Beobachtungen an nachtruhenden Hymenopteren in der umgebung von Mali Lošinj, Kroatien (Anthophoridae, Andrenidae, Eumenidae, Scoliidae, Ichneumonidae). *Joannea Zoologie* 10, 101–121.

Houston TF (1984) Biological observations of bees in the genus *Ctenocolletes* (Hymenoptera: Stenotritidae). *Records of the Western Australian Museum* 11, 153–172.

Kaiser W (1995) Rest at night in some solitary bees – a comparison with the sleep-like state of honey bees. *Apidologie* 26, 213–230.

Keith DA (2004) *Ocean Shores to Desert Dunes: The Native Vegetation of New South Wales and the ACT*. (Department of Environment and Conservation (NSW): Hurstville, NSW)

Linsley EG (1958) The ecology of solitary bees. *Hilgardia* 27, 543–599.

Linsley EG (1962) Sleeping aggregations of aculeate Hymenoptera – I. *Annals of the Entomological Society of America* 55, 148–164.

Linsley EG and Cazier MA (1972) Diurnal and seasonal behavior patterns among adults of *Protoxaea gloriosa* (Hymenoptera, Oxaeidae). *American Museum Novitates* 2509, 1–25.

Mallet J (1986) Gregarious roosting and home range in *Heliconius* butterflies. *National Geographic Research* **2**, 198-215.

Martyn J (2010) *Field Guide to the Bushland of the Lane Cove Valley*. (STEP Inc: Turramurra, NSW)

Marzluff JM, Heinrich B and Marzluff CS (1996) Raven roosts are mobile information centres. *Animal Behaviour* **51**, 89-103.

Mathewson JA and Daly HV (1955) A brief note on the sleep of male *Melissodes* (Hymenoptera: Apidae). *Journal of the Kansas Entomological Society* **28**, 120.

Matthews RW and Matthews JR (2010) *Insect Behavior*. (Springer: New York)

Maynard GV (1991) Revision of *Leioproctus (Protomorpha)* Rayment (Hymenoptera: Colletidae) with description of two new species. *Journal of the Australian Entomological Society* **30**, 67-75.

Maynard G and Rao S (2010) The solitary bee *Leioproctus (Leioproctus) nigrofulvus* (Cockerell 1914) (Hymenoptera: Colletidae), in SE Australia: Unique termite-mound-nesting behavior and impacts of bushfires on local populations. *The Pan-Pacific Entomologist* **86**, 14-19.

Michener CD (1974) *The Social behaviour of the Bees(A Comparative Study)*. (The Belknap Press: Cambridge, Massachusetts)

Miller KK and Wagner MR (1984) Factors influencing pupal distribution of the pandora moth (Lepidoptera: Saturniidae) and their relationship to prescribed burning. *Environmental Entomology* **13**, 430-434.

Moretti M, de Bello F, Roberts SPM and Potts SG (2009) Taxonomical vs. functional responses of bee communities to fire in two contrasting climatic regions. *Journal of Animal Ecology* **78**, 98-108.

Oldroyd BP and Wongsiri S (2006) *Asian Honey Bees: Biology, Conservation and Human Interactions*. (Harvard University Press: Cambridge, Massachusetts)

O'Neill KM (2001) *Solitary Wasps: Behaviour and Natural History*. (Cornell University Press: Ithaca)

O'Toole C and Raw A (1991) *Bees of the World*. (Blandford: London).

Pearson DL and Anderson JJ (1985) Perching heights and nocturnal communal roosts of some tiger beetles (Coleoptera: Cicindelidae) in southeastern Peru. *Biotropica* **17**, 126-129.

Potts SG, Vulliamy B, Dafni A, Ne'eman G, O'Toole C, Roberts S and Willmer P (2003) Response of plant-pollinator communities to fire: changes in diversity, abundance and floral reward structure. *Oikos* **101**, 103-112.

Price OF and Bradstock RA (2010) The effect of fuel age on the spread of fire in sclerophyll forest in the Sydney region of Australia. *International Journal of Wildland Fire* **19**, 35-45.

Rau P (1938) Additional observations on the sleep of insects. *Annals of the Entomological Society of America* **31**, 540-556.

Rau P and Rau N (1916) The sleep of insects; an ecological study. *Annals of the Entomological Society of America* **9**, 227-274.

Raw A (1976) The behaviour of males of the solitary bee *Osmia rufa* (Megachilidae) searching for females. *Behaviour* **56**, 279-285.

Rayment T (1935) *A Cluster of Bees*. (Endeavour Press: Sydney)

Rayment T (1956) The *Nomia australica* Sm. complex: its taxonomy, morphology and biology with the description of a new mutillid wasp. *The Australian Zoologist* **12**, 176-200.

Sapir Y, Shmida A and Ne'eman G (2005) Pollination of *Oncocyclus* irises (*Iris*: Iridaceae) by night-sheltering male bees. *Plant Biology* **7**, 417-424.

Sapir Y, Shmida A and Ne'eman G (2006) Morning floral heat as a reward to the pollinators of *Oncocyclus* irises. *Oecologia* **147**, 53-59.

Schowalter TD (2000) *Insect Ecology: An Ecosystem Approach*. (Academic Press: San Diego).

Schütz S, Weissbecker B, Hummel HE, Apel K-H, Schmitz H and Bleckmann H (1999) Insect antenna as a smoke detector. *Nature* **398**, 298-299.

Silva M De, Andrade-Silva ACR and Silva M (2011) Long-term male aggregations of *Euglossa melanotricha* Moure (Hymenoptera: Apidae) on fern fronds *Serpocaulon triseriale* (Pteridophyta: Polypodiaceae). *Neotropical Entomology* **40**, 548-552.

Stephens PA and Sutherland WJ (1999) Consequences of the Allee effect for behaviour, ecology and conservation. *Trends in Ecology and Evolution* **14**, 401-405.

Stephens PA, Sutherland WJ and Freckleton RP (1999) What is the Allee effect? *Oikos* **87**, 185-190.

Stow A, Silberbauer L, Beattie AJ and Briscoe DA (2007) Fine-scale genetic structure and fire-created habitat patchiness in the Australian Allodapine bee, *Exoneura nigrescens* (Hymenoptera: Apidae). *Journal of Heredity* **98**, 60-66.

Visscher PK, Vetter RS and Robinson GE (1995) Alarm pheromone perception in honey bees is decreased by smoke (Hymenoptera: Apidae). *Journal of Insect Behavior* **8**, 11-18.

Ward P and Zahavi A (1973) The importance of certain assemblages of birds as "information-centres" for food-finding. *Ibis* **115**, 517-534.

Weislo WT (2003) A male sleeping roost of a sweat bee, *Augochlorella neglectula* (CkU.) (Hymenoptera: Halictidae), in Panamá. *Journal of the Kansas Entomological Society* **76**, 55-59.

Webb GA (1994) Sleeping aggregation of insects on *Leptospermum myrtifolium* Sieber ex DC near Bombala, New South Wales. *Victorian Entomologist* **24**, 60-62.

Wertheim B, van Baalen E-JA, Dicke M and Vet LEM (2005) Pheromone-mediated aggregation in nonsocial arthropods: an evolutionary ecological perspective. *Annual Review of Entomology* **50**, 321-346.

Yackel Adams AA (1999) Communal roosting in insects. http://www.colostate.edu/Depts/Entomology/courses/en507/papers_1999/yackel.htm (accessed December 2006).

Received 16 August 2012; accepted 22 November 2012

Koalas *Phascolarctos cinereus* in Framlingham Forest, south-west Victoria: introduction, translocation and the effects of a bushfire

Robert L Wallis

Horsham Campus Research precinct, University of Ballarat, Horsham, Victoria 3402

Abstract

Koalas were introduced into Framlingham Forest, south-west Victoria, in 1971 and the population grew rapidly. By the 1990s the forest was suffering severe defoliation and many trees preferred by Koalas had been over-browsed. In 1998/99 around 1100 Koalas were captured, the males sterilised and animals translocated to other suitable habitats in western Victoria. Some habitat restoration was subsequently undertaken. In 2007 a deliberately lit fire destroyed most eucalypt foliage and many Koalas were killed or burned and removed by wildlife carers and DSE staff. A survey in 2011 found only two Koalas in the area. A Koala management plan for Framlingham Forest has been prepared. (*The Victorian Naturalist* 130 (1) 2013, 37-40)

Keywords: Koala *Phascolarctos cinereus*, management, Framlingham Forest, Indigenous Protected Area.

Introduction

There is debate on the conservation status of Koalas *Phascolarctos cinereus* in Australia, and the Senate Standing Committees on Environment and Communications has recently concluded an inquiry into the species' conservation management and status (Parliament of Australia 2011). In parts of Queensland and NSW there have been dramatic population declines of Koalas mainly through habitat loss, predation and roadkills, and the conservation status has been amended to threatened, while in Victoria and Kangaroo Island, South Australia, the problem facing wildlife managers is over-abundant populations of Koalas that cause severe defoliation, tree death and starvation of the animals.

This paper describes changes in Koala numbers in Framlingham Forest in south-west Victoria from their time of introduction 40 years ago, through a major translocation exercise in the 1990s and after a 2007 bushfire. This forest is a large (1130 ha) remnant native forest that is an Indigenous Protected Area (IPA) – land owned and managed by the local Aboriginal community for cultural and biodiversity conservation.

Framlingham Forest

The forest is located 25 km north-east of Warrnambool in south-west Victoria. The forest is bounded on the east by the Hopkins River and to the north, south and west by cleared farmland. It is a remnant Brown Stringybark *Eucalyptus baxteri* and Manna Gum *E. viminalis* savannah that once dominated the landscape across much of south-west Victoria (Douglas 2004).

Aborigines have been living continuously in and around what is known as Framlingham Forest; in recognition of this, the land was vested by the Victorian Aboriginal Land Act 1987 to the Kirrae Whurrong Aboriginal Corporation. Later it became Victoria's second IPA and the Corporation owns and manages the property for cultural and biodiversity conservation.

Koalas in Framlingham Forest

It is unknown whether Koalas were present in what is now Framlingham Forest before European settlement. Certainly, locals do not recall seeing Koalas in the forest until 1971, when the Victorian Government released 30 unsterilised animals from French Island. There are no details available on the sex of the released animals. These translocated animals were free of the urogenital strain of *Chlamydia* (Martin and Handasyde 1999). Both the Australian and Victorian Governments' Koala Management Strategies list translocation as a method of reducing the impact of overbrowsing by Koalas (DEWHA – Department of Environment, Water, Heritage and the Arts 2009, DSE – Department of Sustainability and Environment 2004).

The population grew rapidly and by the 1990s there were calls for intervention from local naturalists and residents to protect the forest and concerns raised about the health of animals (Martin pers. comm.).

During 1998/1999 DSE began a Koala sterilisation and translocation program. Goldstraw (pers. comm.) estimates some 1100 Koalas were removed to sites that included The Grampians National Park, Mt Cole, Central Highlands and near Casterton. A veterinary surgeon supervised the vasectomies of male animals so that the only fertile males released would theoretically ultimately have been male pouch and 'on-back' young. Koalas were also removed from sites adjoining Framlingham Forest, along the Hopkins River. DSE was unable to provide details of the age, sex and destination of the translocated animals. The overpopulation had destroyed almost all mature Manna Gums and damaged Swamp Gums *E. ovata* and River Red Gums *E. camaldulensis.* Koalas were reported eating non-preferred species such as Messmate Stringbark *E. obliqua* and exotics, which suggests food was limited. A revegetation scheme was undertaken by volunteers (Fig. 1). Goldstraw (pers. comm.) reports that on one occasion he stood in the centre of the picnic ground and counted 11 Koalas in trees surrounding the car park.

Natasha McLean, a postgraduate student at The University of Melbourne, studied the Koalas that were captured and translocated. In 2003 she was awarded a PhD in which, *inter alia*, she examined data for the parameters that contribute to population growth, such as age structure,

sex ratio, and age-specific schedules of mortality and fecundity in a series of overpopulated Koala sites in Victoria (McLean 2003; McLean and Handasyde 2006). The only notable difference between Koalas from Framlingham Forest and those from their original site (French Island) was that at Framlingham Forest 85% of births occurred between December and March, compared with 53% at French Island, indicating a highly seasonal breeding pattern at Framlingham.

On 10 January 2007, a fire that is believed to have been deliberately lit (Thomson and Quirk 2012) raged through the Forest (Warrnambool Wildlife Rescue 2007). Approximately 95% of the remaining Manna Gums were destroyed by this high-intensity fire (Watson pers. comm.) (Figs. 2, 3). DSE records show 147 injured Koalas were rescued by volunteers, although many animals were not reported; a DSE debriefing with veterinarians suggests up to 450 surviving

Fig. 2. Same site post fire.

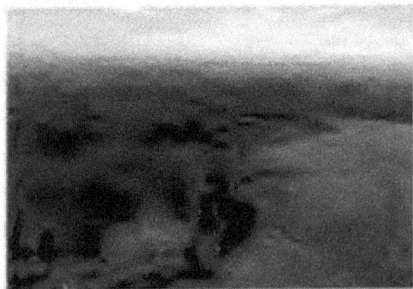

Fig. 3. Vegetation post fire. Some sites (foreground were razed while the canopies in others were badly burned.

Fig. 1. Trees planted by volunteers.

Fig. 4. Aftermath of the January 2007 fire.

Koalas were removed over six weeks post fire. There were reports of many animals killed by the fire, although there are no details available on the actual number (Warrnambool Wildlife Rescue 2007) (Fig. 4).

In 2011 two brief surveys of the road and tracks of Framlingham Forest were undertaken by Deakin University students (9 half hour surveys) and the Warrnambool Field Naturalists Club (2 hours) respectively. Both surveys found two Koalas in the Forest.

Koalas affected by fire

Some 147 Koalas were recorded as rescued after the January 2007 fire. As the responsible agency, DSE coordinated the rescue but members of the Warrnambool Wildlife Rescue group, carers, volunteer wildlife veterinarians and other volunteers undertook much of the rescue operation. Seventy-eight of the collected animals were female; of the animals whose ages were estimated, the 3–5 year old cohort was the most common (23 animals). There were five classed by DSE as 'babies' (presumably pouch young), five animals estimated to be between five months and 10 months old, six classed as over six years old and another four listed as 'adult'.

Animals were collected three days after the fire and for the next 21 days. Most animals were collected 17 days after the fire, although daily collection efforts might have varied.

The fates of 87 animals were recorded; these did not include pouch young that were with

their mothers. Wildlife carers took 38 (the final fate of which is unknown), 33 were released by DSE staff to the nearby Crawford River Regional Park, two went to Healesville Sanctuary, and 14 were either euthanised or died after rescue.

Future management of Koalas in Framlingham Forest

Wallis and Martin (2011) prepared a Koala population management plan for Framlingham Forest for the Kirrae Whurrong Aboriginal Corporation. This report was to inform the Environmental Management Plan for the forest. The Koala population management plan included the following recommendations:

- Regular monitoring of Koalas along a designated survey route should be undertaken. It was recommended regular surveys be conducted and the Koala numbers plotted over time. When rate of growth of numbers appears maximal, DSE Natural Resources – Biodiversity staff should be notified so a population management program can be implemented. Wallis and Martin (2011) suggested this might equate to 50 animals observed along the designated survey route.

- The land managers should not attempt any Koala population regulation activities themselves but instead rely on DSE expertise.

- If the land managers wish to enhance Koala habitat, the health and numbers of *E. viminalis* should be monitored and if these characteristics are deficient after say three years, initiate a planting program of preferred trees, similar to the previous habitat rehabilitation program.

- If Koala numbers again build up, land managers might wish to seek support from the Commonwealth and State Governments to conduct: (i) a eucalypt mapping survey of forest, and (ii) a study of the genetic diversity of the Koala population.

Acknowledgements
I thank Neil Martin, Framlingham Aboriginal Trust, for suggesting the survey work and management plan, for the photographs and for his advice and support throughout the project. I also thank Lisa Shuck and students undertaking ESS420 from Deakin University. Mandy Watson from DSE is especially thanked for providing the data on injured Koalas, her useful comments on the manuscript and the aerial photograph.

Contributions

References

DEWHA - Department of Environment, Water, Heritage and the Arts (2009) *National Koala Conservation and Management Strategy*. (Natural Resource Management Ministerial Council, DEWHA: Canberra) http://www.environment.gov.au/biodiversity/publications/koala-strategy/pubs/koala-strategy.pdf (accessed 7 March 2012).

Department of Sustainability and Environment (2004) *Victoria's Koala Management Strategy*. (DSE: Melbourne)

Douglas J (ed) (2004) *The Nature of Warrnambool* (Warrnambool Field Naturalists Club: Warrnambool)

Martin R and Handasyde K (1999) *The Koala: Natural History, Conservation and Management* (University of NSW Press: Kensington)

McLean N (2003) Ecology and management of overabundant koala (*Phascolarctos cinereus*) populations. (Unpublished PhD thesis, The University of Melbourne)

McLean N and Handasyde K (2006) Sexual maturity, factors affecting the breeding season and breeding in consecutive seasons in populations of overabundant Victorian koalas (*Phascolarctos cinereus*). *Australian Journal of Zoology* 54, 385-392.

Parliament of Australia (2011) *The Koala – Saving our National Icon*. (Senate Committee for Environment and Communications: Canberra) http://www.aph.gov.au/senate/committee/ec_ctte/koalas/report/bo3.htm (accessed 7 March 2012).

Thomson A and Quirk C (2012) Air, ground blitz on Framlingham fire. *Warrnambool Standard* 15 March 2012, p.2.

Wallis R and Martin N (2011) Framlingham Forest Koala Population Management Plan. Unpublished report to Kirrae Whurrong Aboriginal Corporation, Framlingham, Victoria.

Warrnambool Wildlife Rescue (2007) Framlingham bush fire. http://www.warrnambool-wildlife.org.au/gpage7.html (accessed 8 March 2012).

Received 22 March 2012; accepted 26 July 2012

A rare sighting of the Eastern Pygmy-possum *Cercartetus nanus* in north-central Victoria

Anna K Flanagan-Moodie

School of Life and Environmental Sciences, Deakin University, 221 Burwood Highway, Burwood, Victoria 3125
Email: afl@deakin.edu.au

Abstract

The Eastern Pygmy-possum *Cercartetus nanus* is a small, omnivorous marsupial found in south-eastern Australia. In Victoria, the present distribution of *C. nanus* is geographically patchy, generally associated with forests of the Great Dividing Range, but also a range of coastal forests and shrublands. In Box-Ironbark forests of north-central Victoria, *C. nanus* appears to have undergone a severe population decline. In April 2011, a single Eastern Pygmy-possum was observed in the Redcastle-Graytown State Forest during nocturnal survey work. This is a notable record considering that the species has not otherwise been reported from this area for approximately 40 years. (*The Victorian Naturalist* 130 (1) 2013, 40–44)

Keywords: mammal, Box-Ironbark Forest, Redcastle-Graytown State Forest, Heathcote-Graytown National Park.

Introduction

The Eastern Pygmy-possum *Cercartetus nanus* is a small (~17–42 g), omnivorous marsupial of south-eastern Australia (Harris 2008). Its geographic range extends from south-eastern South Australia, through Victoria and New South Wales, to south-eastern Queensland (Ward 1990; Harris *et al* 2007; Menkhorst and Knight 2011); and also includes Tasmania (Harris *et al* 2008). The main threats to the conservation of *C. nanus* are reported to be inappropriate fire regimes resulting in a reduction of the shrub layer, habitat loss, and introduced predators (Harris 2008). The conservation status of the species is listed as vulnerable in New South Wales (New South Wales Government 2012) and South Australia (Government of South Australia 2011). In Victoria, it is not listed as being threatened, although concerns have been expressed about its status (Harris and Goldingay 2005; Harris 2008; Department of Sustainability and Environment 2010).

Cercartetus nanus is considered to be a mid-storey specialist, occurring in a range of vegetation types such as rainforest, sclerophyll forest, shrubland, heathland and woodland (Harris 2008). It is commonly associated with a dense

40

understorey of *Banksia*, or other shrub species, that provide its main food source of nectar and pollen (Menkhorst 1995; Harris 2008). *C. nanus* finds shelter in tree hollows, stumps, *Xanthor-rhoea* skirts and occasionally in disused bird nests (Menkhorst 1995; Harris and Goldingay 2005).

In Victoria, *C. nanus* has a widespread but patchy distribution, occurring in the Portland area in the south-west of the state, the Grampians, the Otway Ranges and the forests of the Great Dividing Range from Ballarat to the north-east of the state (Fig. 1). Its distribution also extends along the south-eastern coastline

from Wilson's Promontory to East Gippsland (Fig. 1) (Menkhorst 1995; Harris and Goldingay 2005).

Recent records of *C. nanus* from the dry Box-Ironbark forests of north-central Victoria have been scarce (Fig. 2), and Menkhorst (1995) suggested that the species may have suffered a severe decline due to a reduction in shrub species associated with intensive management of these forests. The most recent records of *C. nanus* in the Heathcote-Graytown National Park and Redcastle-Graytown State Forest (Fig. 3), east of Bendigo, appear to be prior to 1970 (Fig. 2) (Museum Victoria 2002).

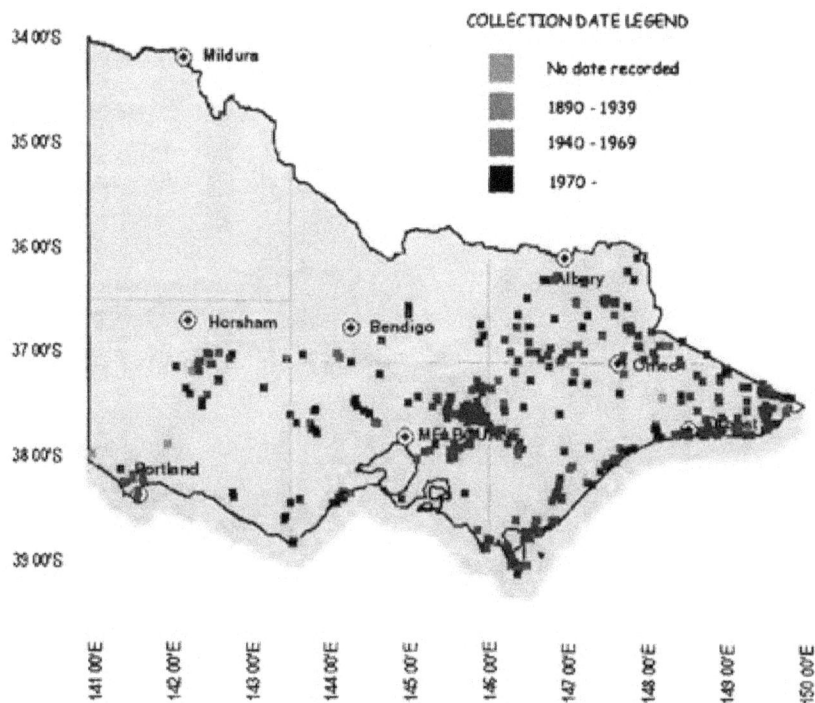

Fig. 1. Distributional records of the Eastern Pygmy-possum *Cercartetus nanus* across Victoria. (source: Museum Victoria 2002).

An observation of *Cercartetus nanus* in Box-Ironbark forest

A study of the ecological effects of prescribed burning in Box-Ironbark forests of the Heathcote-Graytown National Park and Redcastle-Graytown State Forest (an area of approximately 40 000 ha) in north-central Victoria, commenced in 2010 (Bennett *et al.* 2012). This study is examining the effect of prescribed burning on the flora, fauna and structural habitat features, including a study of the Yellow-footed Antechinus *Antechinus flavipes*. As part of the study of *A. flavipes*, approximately 8500 trap-nights of survey effort with small Elliott traps (with peanut butter, oats and honey baits), 7350 camera trap nights with similar baits, and 58 nights of radio telemetry have been performed.

On 12 April 2011, four days after a prescribed burn treatment, two researchers from Deakin University were radio tracking individuals of *A. flavipes* within the recently burnt area. At 8.20 pm, a small animal was spotted on the ground. On closer inspection, the animal started bounding across the ground and climbed to a height of around 1.5 m on a small (2–3 m tall), dead sapling. The animal was observed for approximately 5 minutes, allowing photographs to be taken and identification of the small mammal as *C. nanus* (Fig. 4).

This individual was observed within a study area approximately 100 ha in size within the forest. The overstorey is dominated by Red Ironbark *Eucalyptus tricarpa* L.A.S. Johnson, with Grey Box *E. microcarpa* Maiden, Red Stringybark *E. macrorhyncha* F.Muell. ex Benth.

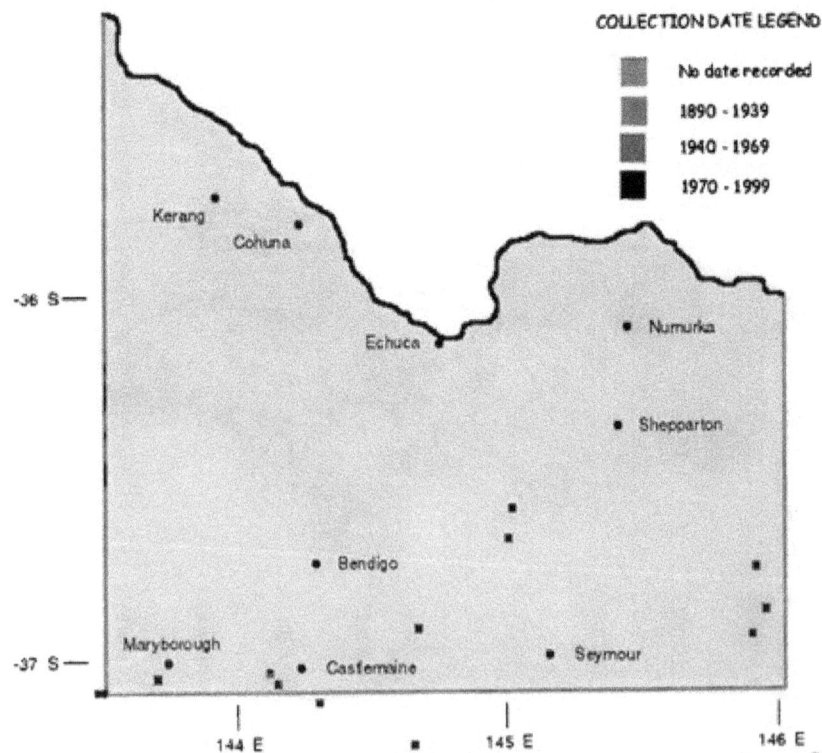

Fig. 2. Distributional records of the Eastern Pygmy-possum *Cercartetus nanus* across north-central Victoria (source: Museum Victoria 2002).

Fig. 3. Regional locality map of north-central Victoria highlighting Heathcote–Graytown National Park, Redcastle–Graytown State Forest and adjoining public land.

Fig. 4. Eastern Pygmy-possum *Cercartetus nanus* sighted in the Redcastle–Graytown State Forest, April 2011. (Photo by Anna Flanagan)

and Red Box *E. polyanthemos* Schauer also present. These overstorey trees have an average diameter of 20–30 cm. Mid-storey and shrub species such as Golden Wattle *Acacia pycnantha* Benth., Grass Tree *Xanthorrhoea glauca* subsp. *angustifolia* D.J.Bedford, Sweet Bursaria *Bursaria spinosa* Cav. and Drooping Cassinia *Cassinia arcuata* R.Br. also occur in this area.

Conclusion

This sighting of a single *C. nanus* in Redcastle-Graytown State Forest is significant as it confirms the continued presence of the species in this area, despite it not having been recorded here for at least 40 years. The lack of other recent records during the current surveys, despite extensive small mammal trapping, camera trapping and radio telemetry undertaken in the area, suggests that the size of the present population of *C. nanus* is small.

Other methods for detecting this species, such as the use of nest boxes, pitfall trapping and analysis of predator scats (Bennett *et al.* 1989; Bladon *et al.* 2002; Harris and Goldingay 2005; Harris 2008), may also be useful in gaining additional records in these forests. The deployment of nest boxes developed particularly for this species may be a useful management

Contributions

strategy to assist with recovery and monitoring of *C. nanus* in this area (e.g. Bladon *et al.* 2002).

The confirmation of the ongoing presence of *C. nanus* in these Box-Ironbark forests of central Victoria is noteworthy, and needs to be considered in developing wildlife management strategies for the region.

Acknowledgements

Thanks to Andrew Bennett, Greg Holland and Mike Clarke for their input to the management of the fire ecology study, and for comments on this article. Financial support towards the fire ecology study has been provided by the Department of Sustainability and Environment (NW Region and Project Hawkeye), Parks Victoria, Deakin University, La Trobe University, and the Holsworth Wildlife Foundation. Thanks also to Alicia Ivory for her assistance in the field. This work has been undertaken under Flora and Fauna permit number 10005470.

References

Bennett AF, Holland GJ, Flanagan A, Kelly S and Clarke MF (2012) Fire and its interaction with ecological processes in Box-Ironbark forests. *Proceedings of the Royal Society of Victoria.* **124**, 72-78

Bennett AF, Schulz M, Lumsden LF, Robertson P and Johnson PG (1989) Pitfall trapping for small mammals in temperate forest environments. *Australian Mammalogy* **12**, 37-39.

Bladon RV, Dickman CR and Hume ID (2002) Effects of habitat fragmentation on the demography, movements and social organisation of the eastern pygmy possum (*Cercartetus nanus*) in northern New South Wales. *Wildlife Research* **29**, 105-116.

Department of Sustainability and Environment (2010) Flora and Fauna Guarantee Act 1988 - Threatened List - October 2010. Melbourne.

Government of South Australia (2011) *National Parks and Wildlife Act 1972*

Harris J (2008) *Cercartetus nanus* (Diprotodontia: Burramyidae). *Mammalian Species* **815**, 1-10.

Harris JM and Goldingay RL (2005) Distribution, habitat and conservation status of the eastern pygmy-possum *Cercartetus nanus* in Victoria. *Australian Mammalogy* **27**, 185-210.

Harris JM, Gynther IC, Eyre T, Goldingay RL and Mathieson MT (2007) Distribution, habitat and conservation status of the eastern pygmy-possum *Cercartetus nanus* in Queensland. *Australian Zoologist* **34**, 209-216.

Harris JM, Munks SA, Goldingay RL, Wapstra M and Hird D (2008) Distribution, habitat and conservation status of the eastern pygmy-possum *Cercartetus nanus* in Tasmania. *Australian Mammalogy* **29**, 213-232

Menkhorst PW (ed) (1995) *Mammals of Victoria : distribution, ecology and conservation* (Oxford University Press in association with Department of Conservation and Natural Resources, Melbourne)

Menkhorst PW and Knight F (2011) *A field guide to the mammals of Australia* (Oxford University Press: South Melbourne)

Museum Victoria (2002) *Bioinformatics Victorian Faunal Web Site.* Published on the Internet; http://www.museum.vic.gov.au/bioinformatics/ [accessed 19 March 2012 at 16:30; search string: *Cercartetus nanus*]. Melbourne, Australia.

Parliament of New South Wales (2012) *Threatened Species Conservation Act 1995*

Ward SJ (1990) Life history of the Eastern Pygmy-possum. *Cercartetus nanus* (Burramyidae: Marsupialia), in south-eastern Australia. *Australian Journal of Zoology* **38**, 287-304.

Received 17 May 2012; accepted 11 October 2012

Ninety-seven Years Ago

'POSSUMS. -- "F.R." in "Bush Notes" in the *Australasian* of 24th June, has some remarks on the food of 'possums in captivity. He says that in their wild state it was obviously impossible for 'possums to obtain cooked meat, yet in confinement they will eat it freely. Again, before the advent of the white man fruit was practically non-existent in Australian forests; yet a neighbour, who has a choice garden containing some fine apple-trees, finds that as soon as the apples begin to ripen the 'possums begin to arrive, though no one would suspect that there were any of the animals in the neighbourhood. They are very fond of apples, and will also eat peaches and other fruits, while potatoes and other vegetables are also favoured. In another friend's garden the buds of a La France rose were continually disappearing, the cause being put down to snails, but it was afterwards found that 'possums were the cause, and "F.R." says in his own garden they will leave everything else for a rosebud. Cooked meat, he remarks, seems to have irresistible attractions for many wild things. Cockatoos and parrots are very fond of it, yet, of course, they could not possibly have tasted it in their wild state. There is no accounting for these aberrations, and apparently these strange articles of diet do them no harm—in fact, they seem to thrive on them. Cake and sugar are common articles of diet with tame 'possums. These must be very different from the meals of gum-leaves that formed their natural food. No doubt they also eat grass in their native state, but their staple food is undoubtedly the young shoots and leaves of the various eucalypts.

From *The Victorian Naturalist* **XXXIII**, pp. 47-48, July 6, 1916

The rare collembolan genus, *Temeritas* (Symphypleona: Sminthuridae), in southern Australia: systematics, distribution and conservation status

Penelope Greenslade

Centre for Environmental Management, University of Ballarat, Ballarat, Victoria 3353
South Australian Museum, North Terrace, Adelaide, South Australia 5000.
School of Biology, Australian National University, Australian Capital Territory 0200.
Email: Pgreenslade@staff.ballarat.edu.au.

Abstract
A brief summary of the genus *Temeritas* is given with distributions of the four described Australian species and records of other species in the genus. A spelling correction is documented for the Western Australian species and a new name for the Victorian species is formally proposed here as the original name is preoccupied. Characters that distinguish *Temeritas* from allied genera are noted and the conservation status of the three southern species and Collembola in general are discussed. (*The Victorian Naturalist* 130 (1) 2013, 45–48)

Keywords: *Temeritas denisi, Temeritas regalis, Temeritas elegans*

Introduction
The Collembola, common name Springtail, are a group of arthropods, little known because of their small size and cryptic habits. However, the class is abundant, widespread and species-rich with a high proportion of species endemic to Australia. Genera in some families include a high proportion of endemic species (short-range endemics), an example being the genus *Temeritas* Delamare Deboutteville and Massoud, 1963.

Species of *Temeritas* are easily recognised as they are globular, up to 2 mm long, usually brightly coloured with purple stripes and spots and slender antennae that are longer than the head and body combined (Figs 1, 2, 3).

The genus has a predominantly pantropical distribution although, exceptionally, some species in Australia and New Zealand are restricted to temperate climates. At present, 47 species are known in the genus, of which three have been described from Australia (Bellinger *et al.* 2012). The Australian species were described originally in the genus *Sminthurus* Latreille as *S. denisi* Womersley, 1934 from south-west Western Australia, *S. regalis* Womersley, 1939 from southern South Australia and *S. elegans* Womersley, 1939 from southern Victoria. These species are clearly allopatric. *Sminthurus denisi* was incorrectly named as *S. denisii* by Womersley (1934) but, in his subsequent publications

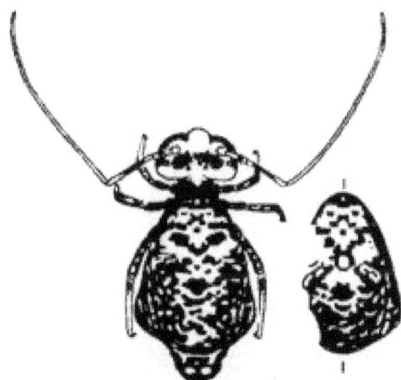

Fig. 1. Pen and ink drawing of *Temeritas regalis* (Womersley) by JM Betsch

(1936, 1939), he correctly changed the specific name to *S. denisi*. Womersley also recorded this species from New Zealand (Womersley 1936) but the record is unlikely and remains unconfirmed. Najt (1968) transferred two of the Australian species, *S. regalis* and *S. elegans*, to the genus *Temeritas*, but transposed their localities in her publication. Later, *S. denisi* was also transferred to *Temeritas* by Greenslade (1994).

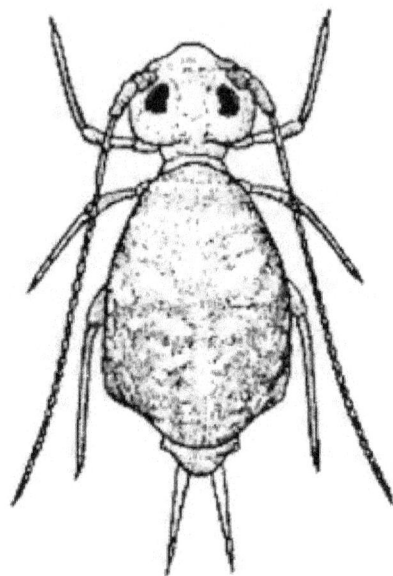

Fig 2. Line drawing of *Temeritas denisi* (Womersley) from Womersley (1939)

Fig. 3. Line drawing of *Temeritas isabellae* Greenslade from Womersley (1939)

Distribution

A number of undescribed species have been collected from the northern regions of Western Australia, Northern Territory and Queensland including the Torres Strait Islands. The species from Murray Island, Torres Strait, has been identified as *Temeritas womersleyi* (Denis,

1948) originally described from Vietnam (P. Greenslade new record). Localities from where undescribed species have been collected in the last 20 years are Sweers Island, Kuranda, Woodstock and Lamington National Park in Queensland, McArthur River and Jabiru in the Northern Territory, New England National Park, Lord Howe Island and Norfolk Island in New South Wales, and Brookton and Barrow Island in Western Australia. The genus has not been collected from Tasmania.

Systematics

Among the other genera of globular Collembola in Australia, *Temeritas* is distinct because of the relatively large size of species (nearly 2 mm long), their strongly annulated antennae that are longer than the body, and attractive colour patterns and bands (Fig. 1). Species of *Temeritas* can be distinguished from *Sminthurus*, in which genus Womersley (1934) first described the southern Australian species, by the lack of an internal spine on trochanter III and the shorter antennae. There is only one species of *Sminthurus* in Australia, the pest species *S. viridis* (L.). There are three other genera in Australia with which *Temeritas* might be confused but are distinct because of their long, strongly anulated antennae, often relatively large size and attractive colour patterns, particularly in the banded antennae (Fig. 4). The first of these is *Parropalites* Bonet and Tellez, new record, which is known only from rainforests in northern Queensland; but this genus is poorly characterised at present and it has much shorter antennae. The second genus with which *Temeritas* could be confused is *Pararrhopalites* Bonet and Tellez, new record. The third is *Sphyrotheca* Börner, another superficially similar genus but it has a neosminthuroid seta ventrolaterally on abdominal segment IV and also antennae that are shorter than the body.

Nomenclature

As the original name of the Victorian *Sminthurus elegans* is preoccupied by *Sminthurus elegans* Fitch, 1863 from North America (now *Sminthurinus elegans*), even though the Australian species has been transferred to a different genus, a new name is required and it is renamed here as *Temeritas isabellae*.

Table 1. Total number of described species, endemic species and percentage of endemic species in each family of Collembola known from Australia

Higher taxon	Total described species	Number of endemic species	% endemic species
Arrhopalitidae	2	0	0
Bourletiellidae	24	16	66
Brachystomellidae	21	18	85
Cyphoderidae	4	2	50
Dicyrtomidae	4	3	75
Entomobryidae	76	33	50
Hypogastruridae	28	7	25
Isotomidae	53	23	44
Katiannidae	32	24	75
Neanuridae	46	36	78
Neelidae	3	0	0
Odontellidae	4	3	75
Oncopoduridae	1	1	100
Onychiuridae	5	0	0
Paronellidae	14	10	71
Sminthuridae	6	3	50
Sminthurididae	6	2	33
Spinothecidae	1	1	100
Tomoceridae	7	3	43
Tullbergiidae	17	8	47
Mean			53

Conservation status

In spite of the high proportion of nationally endemic and locally endemic species, the conservation of only a few Collembola has been given attention. *Tasphorura vesiculata* Greenslade and Rusek (Tullbergiidae), occurs only in moss in a small patch of rainforest in north-east Tasmania, which has been given a low level of protection in that any future logging must consider protecting this species. Another is a species of *Australotomurus*, Stach (Entomobryidae: Orchesellinae) found in only four vegetation remnants in urban Perth. It was listed with the *Western Australian Wildlife Conservation Act* 1950 but delisted a few years later even though one of the remnants was partially alienated (P. Greenslade submitted). Some other genera of Collembola known to contain short range endemics are *Nasosminthurus* Stach (Bourle-

tiellidae), *Epimetrura* Schött (Entomobryidae) and *Folsomotoma* Bagnall (Isotomidae). The percentages of endemic compared with total species in each family known from Australia are given in Table 1. Total mean endemism is 53% and several of the larger families contain more than 70% endemic species (Brachystomellidae, Katiannidae, Neanuridae, Paronellidae) (Table 1). Many of these endemic species would be short range endemics.

All southern Australian species of *Temeritas* are uncommon and patchy in distribution. Because their habitats include leaf litter, native grasses, moss and under logs in humid forests, they are likely to be susceptible to climate warming. Indeed, some populations may already have become locally extinct as a result of drought and competition from invasive exotic species as well as longer term climate

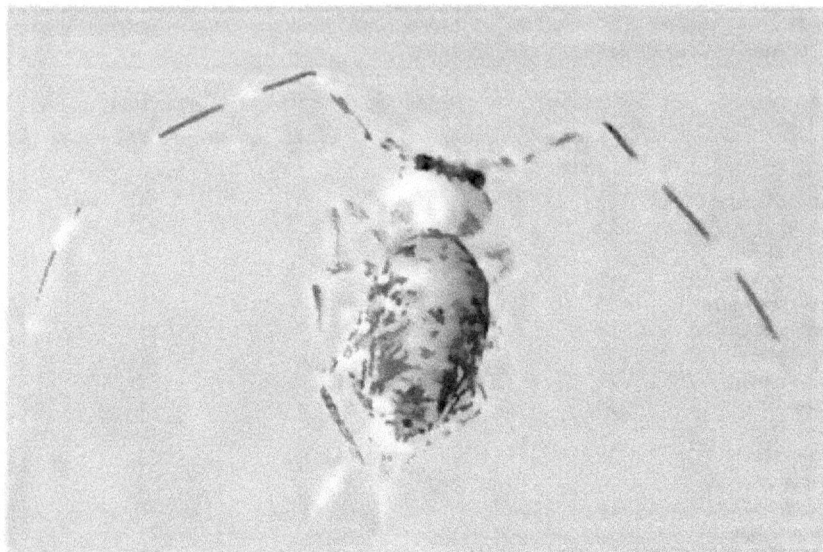

Fig. 4. Photograph of *Temeritus isabellae* from the Dandenong Ranges, Victoria.

change. For instance, *T. regalis* was collected relatively frequently in the southern Mt Lofty Ranges in the 1970s but has not been found in the last 20 years in localities where it was previously present. *Temeritas isabellae* was described originally from Kalorama, Mt Dandenong and there are also old records from Erskine River, Toolangi State Forest, Bellel Creek, Coranderrk Reserve and Silverband Falls in the Grampians, all in Victoria. As with *T. regalis*, the Victorian species has not been collected in the last 30 years. A search in suitable localities and habitats should be undertaken as a priority, to establish its current conservation status.

References
Bellinger PF, Christiansen KA and Janssens F (1996–2012) Checklist of the Collembola of the World. http://www. collembola. org Accessed June 2012.

Greenslade P (1994) Collembola. In *Zoological Catalogue of Australia*, pp. 19–138; 157–184. Ed WWK Houston. Volume 22: Protura, Collembola, Diplura. (CSIRO: Collingwood, Vic)

Najt J (1968) Nouveaux documents sur le genre *Temeritas* et sa distribution géographique (Collembole Symphypléone). *Revue d'Écologie et Biologie du Sol* 5, 631–636.

Womersley H (1933) On some additions to the sminthurid fauna of Australia. *Stylops* 2, 241–247.

Womersley H (1934) Notes on some Australian Collembola. *Stylops* 3, 244–246.

Womersley H (1936) On the Collembolan fauna of New Zealand. *Transactions of the Royal Society of New Zealand* 66, 316–328.

Womersley H (1939) *Primitive Insects of South Australia. Silverfish, Springtails and their allies.* (Government Printer: Adelaide)

Received 28 June 2012; accepted 30 August 2012

Leech predation of frog spawn

Introduction

The predators of Australian anurans and their larvae are well documented (Tyler 1976, 1994, Littlejohn and Wainer 1978; Davies *et al.* 1979; Morgan and Buttemer 1996, Gillespie and Hero 1999). By contrast, little has been published on the sources of predation of their spawn. Tyler (1976, 1994) states that there are relatively few predators of frog spawn and that 'fish probably constitute the major predator'. He notes in particular that the foam nests of the genus *Limnodynastes* are probably most accessible to terrestrial insects because they tend to be located around the edges of ponds where they are attached to peripheral vegetation, and that they are occasionally eaten by ants. Members of the Australian frog genus *Limnodynastes* produce floating foam-capped nests below which the egg mass resides (Parker 1940; Tyler and Davies 1979; Roberts 1989). One member of this genus, the Spotted Marsh Frog *Limnodynastes tasmaniensis*, is a very common species throughout much of south-eastern Australia where it breeds in most months of the year in both temporary and permanent water bodies and in a wide variety of both natural and manmade habitats (Barker *et al.* 1995; Hero *et al.* 1991; Littlejohn 2003). Herein I report the predation of *L. tasmaniensis* spawn by leeches in an ephemeral wetland near Melbourne some 25 years ago and compare these observations with a very similar report of predation documented by Burgin and Schell (2005) in the Sydney area.

Observations

1. On 6 January 1987, following two days of heavy rain, a shallow ephemeral wetland located in remnant River Red Gum *Eucalyptus camaldulensis* woodland adjacent to the Darebin Creek in the north of Bundoora (37°69'S, 145°05'E) Victoria, was visited. The swamp had been completely dry since about mid-December of the previous year but rain had refilled it and had stimulated a burst of breeding activity in *L. tasmaniensis*. There were large persistent daytime choruses (> 50 males) and numerous freshly deposited foam nests around clumps of aquatic vegetation. Most nests were aggregated amongst a 9 m² patch of Spikerush *Eleocharis sphaecelata* where they were exposed to dappled sunlight or else were completely shaded. A total of 27 nests were located in this patch. The site was visited over four consecutive days and nests inspected for the presence of leeches and other invertebrates on each occasion. Water temperature approximately 10 cm below the surface varied between 21-24°C at midday over the four days.

Leeches were observed on the foam caps of *L. tasmaniensis* nests on each day. All of the leeches appeared to belong to the same species and were uniform black in colour and approx. 50-60 mm in length. (Leeches were not able to be identified to genus (or species) level owing to the lack of an appropriate identification guide at the time.) The leeches were observed typically lying completely still on the foam cap of the nests with the head and anterior body buried down through the foam cap into the gelatinous egg mass below. While most of the affected nests contained a single leech, on three nests there were two, and on one nest, three leeches. On three nests the surface of the foam caps had dried to a polystyrene-like consistency and leeches had attached themselves to the side of the nest where they were just visible above the water line. Nests around the periphery of the aggregation were most affected by leeches while only one leech was recorded on a nest near the 'centre'. On the first day, three of the leeches (taken from nests outside of the aggregation) were euthanised and found to contain numerous (> 10), mostly intact frog's eggs.

The incidence of leeches on foam nests remained fairly constant over the four days, affecting about one-third of all nests (30-37%;

Table 1). By the fourth day, the eggs of seven nests had begun to hatch and most of the others were close to hatching (Gosner stages 20- 25; Duellman and Trueb 1986). Two nests were occupied by leeches for up to three consecutive days. Eight leeches closely examined on the fourth day had noticeably distended bodies, indicating recent feeding.

Other arthropods located on the foam caps of nests included (total number in parenthesis): ants (6), aquatic snails (5), spiders (4), caterpillars (3), millipedes (2) and dipterans (8). As none of these arthropods appeared to be feeding directly on the eggs, it is likely that these occurrences were quite incidental and represent fauna displaced by flooding (although see Discussion). The percentage of nests with other arthropods was consistent over the three days they were recorded (14-16%; Table 1).

In addition to the observations above, I have since made very similar observations at two other (nearby) sites:

2. Approximately 3 km south of the above site, beside the Darebin Creek in Bundoora, two leeches were located on separate, recently deposited *L. tasmaniensis* nests in a relatively small ephemeral pond following rain in January.
3. At Somerton (37°63'S, 144°95'E) near the southern boundary of Craigieburn Grasslands, four leeches were located separately on the foam caps of freshly laid *L. tasmaniensis* nests partially concealed by *Poa* sp. tussocks and deposited in a large ephemeral pond which had been filled by heavy rain in November.

At all three localities the leeches found on *L. tasmaniensis* nests appeared to be the same species. These leeches were occasionally caught in dip-nets skimmed through water around the periphery of large ponds and swamps at the sites, indicating their aquatic habit. While *L. tasmaniensis* has frequently been observed to breed in small ephemeral ponds (n > 15), no leeches were ever observed on nests deposited in these ponds. Leeches were never observed as ectoparasites of *L. tasmaniensis* larvae or adult frogs at any of the sites, despite regular visits over more than ten years.

Discussion

The sanguivorous habit of many terrestrial and aquatic leeches is well known and leeches have been documented as ecto and endoparasites of both frogs and their larvae (Waite 1925; Mann and Tyler 1963; Brockelman 1969; Tyler 1976; Duellman and Trueb 1986; Sawyer 1986 and references therein; McCallum *et al.* 2011). By contrast the literature on leeches as macrophagous predators of frog spawn, though relatively small, has been largely neglected or omitted entirely from consideration in reviews of both leech and amphibian biology (Duellman and Trueb 1986; Govedich 2001; Toledo 2005; Romano and Di Cerbo 2007). A relatively recent literature review by Romano and Di Cerbo (2007) found that anuran egg predation by leeches had been documented in some 20 species, representing 3.6% of the total number of anuran species in those regions where anuran leech predation occurred. That some leech species should consume frog spawn is curious

Table 1. The frequency of occurrence of leeches and other arthropods on 27 foam nests of the Spotted Marsh Frog *Limnodynastes tasmaniensis* monitored over a four day period.

Day	Number of Leeches	% of nests with Leeches	% of nests with other arthropods
1	11	30	15
2	15	37	16
3	16	37	14
4	10	33	–

because it occurs in spite of a clear adaptation they have to piercing the skin of mammals (Cargo 1960) and other vertebrates. *Limnodynastes tasmaniensis* is the only Australian frog species in which this kind of predation has been documented to date.

The presence of leeches on *L. tasmaniensis* nests is unlikely to be the result of their displacement due to flooding for two reasons: (i) I had only ever located them in water and thus their presence on the top or sides of foam nests above the water level (in many instances) is at odds with this habit, and (ii) in all instances the head of the leech was protruding down through the foam cap into the egg mass, consistent with their being engaged in feeding. Even if the leeches were present on foam nests due to disturbance of some kind, the small sample of leeches found to have consumed frog spawn indicates opportunistic feeding was occurring. The number of leeches recorded on individual foam nests in this work must, however, be considered an underestimate as only a few nests were thoroughly examined for leeches residing amongst the egg mass or the portion of the egg mass below the water (and none were located).

The impact that the leeches had on individual nests was not apparently severe since their presence did not seem to affect the integrity of the nests and the relatively warm conditions meant that egg development was rapid, ensuring that most eggs hatched to produce larvae.

The occurrence of dipterans on nests, while possibly incidental, is worthy of closer examination as the parasitisation of frog spawn by dipteran larvae has been documented to occur in various other anuran species (Bokermann 1957; Tyler 1976; Villa *et al.* 1982; Menin and Giaretta 2003). Furthermore six South American leptodactylidae frog species (that produce foam nests similar to *L. tasmaniensis*) were found to suffer significant predation from dipteran larvae (Menin and Giaretta 2003).

It seems remarkable, given how common *L. tasmaniensis* is in south-eastern Australia, and the conspicuousness of black leeches on the contrasting white foam nests, that leech predation had not been reported until relatively recently. This may indicate that leech predation does not occur in all breeding situations, or is limited by the distribution and/or habitat

preferences of the particular species of leech involved.

Burgin and Schell (2005) reported the leech *Bassianobdella fusca* feeding on *L. tasmaniensis* foam nests from a wetland near Sydney and most of the observations described above are consistent with their work. For instance, the timing of the observations in both cases was summer (or late spring), coinciding with maximum leech activity, and both sets of observations occurred in large ephemeral water bodies. One notable point of difference was that Burgin and Schell (*op. cit.*) observed that leeches consumed ova only in Gosner stages 1–14, which meant that clutches were vulnerable to predation only in the first 24 hours following oviposition; observations in this work indicate that leeches remained on spawn clumps, apparently continuing to feed, for up to four days. It would be useful to know if this same leech species was also responsible for predation events described in this work, and further, whether leeches are able to consume larger and more developmentally advanced larvae (i.e. Gosner stages > 14).

Finally, Håkansson and Loman (2004) have shown that spawn located in the centre of communal aggregations of the Common Frog *Rana temporaria* suffered markedly less leech predation compared to those on the periphery. A similar pattern of leech predation was noted in this work and may be worthy of more detailed examination.

References

Barker J, Grigg GC and Tyler MJ (1995) *A Field Guide to Australian Frogs*. (Surrey Beatty & Sons: Chipping Norton, NSW)

Bokermann WCA (1957) Frog eggs parasitized by dipterous larvae. *Herpetologica* 13, 231–232.

Brockelman WY (1969) An analysis of density effects and predation in *Bufo americanus* tadpoles. *Ecology* 50, 632–644.

Burgin S and Schell CB (2005) Frog eggs: unique food source for the leech *Bassianobdella fusca*. *Acta Zoologica Sinica* 51, 349–353.

Cargo DG (1960) Predation of eggs of the spotted salamander, *Ambystoma maculatum*, by the leech *Macrobdella decora*. *Chesapeake Science* 1(3), 119–120.

Davies M, Tyler MJ and Martin AA (1979) Frogs Preyed on by Ants? *The Victorian Naturalist* 96(3), 97.

Duellman W E and Trueb L (1986) *Biology of Amphibians*. (McGraw-Hill: New York)

Gillespie, GR and Hero, J-M (1999) Potential impacts of introduced fish and fish translocations on Australian amphibians. In *Declines and Disappearances of Australian Frogs*, pp. 137–145. Ed A Campbell. (Environment Australia: Canberra)

Govedich FR (2001) *A Reference Guide to the Ecology and Taxonomy of Freshwater and Terrestrial Leeches (Euhirudinea) of Australasia and Oceania.* (Cooperative Research Centre for Freshwater Ecology. Identification Guide No. 35: Thurgoona, NSW)

Håkansson P and Loman J (2004) Communal spawning in the Common Frog *Rana temporaria* – Egg temperature and predation consequences. *Ethology* 110, 665–680.

Hero JM, Littlejohn M and Maraatelli G (1991) *Frogwatch Field Guide to Victorian Frogs.* (Department of Conservation & Environment: Melbourne)

Littlejohn MJ (2003) *Frogs of Tasmania. Fauna of Tasmania.* Handbook No. 6. 2 edn. (University of Tasmania: Hobart)

Littlejohn MJ and Wainer JW (1978) Carabid beetle preying on frogs. *The Victorian Naturalist* 95, 251–252.

Mann KH and Tyler MJ (1963) Leeches as endoparasites of frogs. *Nature* (London) 197, 1224–1225.

McCallum ML, Moser WE, Wheeler BA and Trauth SE (2011) Amphibian infestation and host size preference by the leech *Placobdella picta* (Verrill, 1872) (Hirudinida: Rhynchobdellida: Glossiphoniidae) from the Eastern Ozarks, USA. *Herpetology Notes* 4, 147–151.

Menin M and Giaretta AA (2003) Predation on foam nests of leptodactyline frogs (Anura: Leptodactylidae) by larvae of *Beckeriella niger* (Diptera: Ephydridae). *Journal of Zoology* (London) 261, 239–243.

Morgan LA and Buttemer WA (1996) Predation by the non-native fish *Gambusia holbrooki* on small *Litoria aurea* and *L. dentata* tadpoles. *Australian Zoologist* 30(2), 143–149.

Parker HW (1940) The Australasian frogs of the family Leptodactylidae. *Novitates Zoologicae* 42, 1–106.

Romano A and Di Cerbo AR (2007) Leech predation on Amphibian eggs. *Acta Zoologica Sinica* 53, 750–754.

Sawyer RT (1986) *Leech Biology and Behavior.* Volumes I, II & III. (Clarendon Press: Oxford)

Toledo, LF (2005) Predation of juvenile and adult anurans by invertebrates: current knowledge and perspectives. *Herpetological Reviews* 36, 395–400.

Tyler MJ (1976) *Frogs* (Collins: Sydney)

Tyler MJ (1994) *Australian Frogs – a natural history.* (Reed Books: Chatswood, NSW)

Tyler MJ and Davies M (1979) Foam nest construction by Australian Leptodactylid Frogs (Amphibia, Anura, Leptodactylidae). *Journal of Herpetology* 13, 509–510.

Villa J, McDiarmid RW and Gallardo JM (1982) Arthropod predators of leptodactylid frog foam nests. *Brenesia* 19/20, 577–589.

Waite ER (1925) Field notes on some Australian reptiles and a batrachian. *Records of the South Australian Museum* 3, 17–32.

Grant S Turner
103 Settlement Road
Bundoora, Victoria 3083

One Hundred and Twenty-two Years Ago

Notes On The Planarian Worms Obtained On The Upper Wellington.

BY ARTHUR DENDY

1. *Geoplana howitti*, species nova.—Unfortunately only a single specimen of this worm was found, but it is a well marked and very beautiful species. The ground colour of the dorsal surface is yellowish white. In the middle line there is a fairly broad band of the ground colour, and on each side of this a stripe of about equal width of dark purplish brown, then a rather broader band of ground colour thickly flecked with dark purplish brown and edged on the outside by a fine line of the same. Outside this is a very narrow margin of ground colour. All the dark bands unite at each end. The ventral surface is pale yellowish white or grey, with no markings.

2. *Geoplana lucasi*, Dendy.—This is a remarkable and very rare planarian, of unusually large size, and with black and white markings. It was hitherto known only from three specimens found on the top of the coast ranges in the Croajingolong district, on the occasion of the Club's expedition to that locality, and described (from spirit specimens only) by me in the " Transactions of the Royal Society of Victoria." Only a single specimen was found.

3. *Geoplana quadrangulata*, Dendy.—A small variety of this remarkable species was found in abundance. Hitherto it has only been recorded from Macedon, and in very small numbers.

4. *Geoplana frosti*, Spencer.—This species was recently discovered on the Club's expedition to the Yarra Falls, and is described by Professor Spencer in the "Transactions of the Royal Society of Victoria." We obtained one small specimen.

5. *Geoplana alba*, Dendy.—We obtained several fine examples of this common planarian.

6. *Geoplana sulphurea*, Fletcher and Hamilton.—This species was common.

From *The Victorian Naturalist*, VIII, pp. 43–44, June – July, 1891

Dorothy Mahler

28 February 1941 – 12 December 2012

Dorothy Mahler was elected to the FNCV in 1985 and within a few years began an active contribution to the operation of the Club that was to continue until the end of 2012. In that time she occupied a range of positions within the Club.

Dorothy's natural history subject of choice was birdlife but she had wide ranging interests and was an active member of both the Botany Group, of which she was Assistant Secretary in 1989, and the Geology Group. She contributed reports on excursions undertaken by these groups as well as occasional reports of meetings of the Fauna Survey Group. On two occasions, having taken part in the Annual Camp of the Victorian Field Naturalists Clubs Association, she provided reports that detailed the activities that took place.

For more than 20 years Dorothy was involved in most aspects of production of the Club's newsletter, Field Nats News. From its first issue, in November 1990, until the December 1992/January 1993 issue, it was Dorothy who typed up and laid out the material, in her spare time at her work place. She would then deliver it to a nearby 'Pink Panther', for printing. In the early days of Field Nats News, Dorothy and Noel Schleiger between them also did all the collation of the 500 plus copies of the newsletter. Dorothy continued to assist with collation well past issue no. 200.

As well as her role with the newsletter, Dorothy contributed to the smooth operation of the Club's journal. From 2001 until October 2012 Dorothy undertook the essential task of sending complimentary copies of *The Victorian Naturalist* to authors, following the publication of each issue.

In June 1990 Dorothy took on the role of Excursion Secretary. Until she stepped down in May 1998 she organised, was the contact point for, took part in (often as leader), and reported on dozens of FNCV general excursions. All parts of the metropolitan area, as well as locations within easy driving distance of Mel-

bourne, were covered. During this period Dorothy was also the FNCV Tour Operator, and organised what became memorable interstate trips. Destinations included Binna Burra, within Lamington National Park in Queensland (August 1991); the northern coast of Tasmania (11–24 January 1992); Kangaroo Island (10–23 October 1993); Lake Mungo and Mootwingee (26 August–7 September 1995) and the Mount Kosciusko area (17–25 January 1997). Extended trips were planned and undertaken also to the Grampians (six days in October 1992) and south-western Victoria (three days in March 1993).

In 1994, the same time that she was organising some of these activities, Dorothy also served as a member of FNCV Council.

Dorothy was made an Honorary Member of FNCV in August 2012, along with her partner, Noel Schleiger, for their individual and joint contributions to the Club. Dorothy's input to FNCV was wide-ranging, significant and enduring. As Valda Dedman wrote in 2005 ('The

Victorian Naturalist 122: 309) 'Dorothy Mahler is a great worker ... She represents the indispensable 'backroom girls', not on Council, but essential to the Club.'

Gary Presland
40 William Street
Box Hill 3128

Wetland Weeds: Causes, Cures and Compromises

by Nick Romanowski

Publisher: CSIRO Publishing, Collingwood, Victoria 2011. 140 pages, paperback, colour photographs. ISBN 9780643103955. RRP $49.95.

Nick Romanowski has been infatuated with indigenous wetland plants for over four decades and his passion shows in *Wetland Weeds: Causes, Cures and Compromises*, a book written in his endeavours to educate people about the dangers of using introduced plants in aquaria and ponds or water gardens. His efforts are commendable. Weeds of waterways have many costs — environmental, economic and cultural. They can out-compete desired native plants, thus reducing biodiversity; form dense infestations that clog waterways, making their navigation difficult and impeding recreational activities, irrigation and industrial processes; and divert waterflow, resulting in erosion and/or flooding. Weeds also can be difficult and costly to eradicate. There is much information concerning the problems caused by aquatic weeds (e.g. Adair and Groves 1998; Groves *et al.* 2005), as well as examples of the cost of their management, such as $1.6 million for the *Salvinia* infestation in the Hawkesbury-Nepean River in 2004 (Gorham 2008) and $140000 per annum for *Cabomba caroliniana* Fanwort in Lake Macdonald in the Noosa biosphere in Queensland (Moran 2009). As Romanowski says (page 13), the primary theoretical defence against weeds is education. I would have preferred, therefore, that the sections in Chapter 1 on problems caused by weeds and the causes of weediness, had provid-

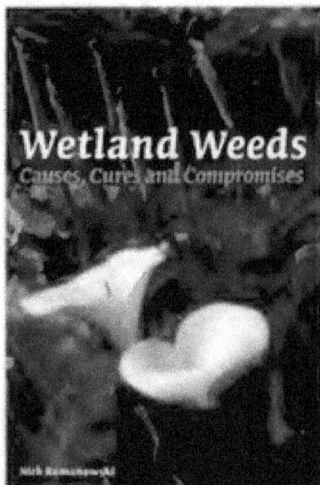

ed much greater detail. The chapter, however, does provide a good overview of what a weed is and the legal and official categories of weeds.

Chapter 2 discusses prevention, different types of control and management of weeds. The author pragmatically explains the importance of differentiating between the various types and levels of threat posed by weeds and the likelihood of eradication or control. At times, however, he tends to ramble and rely on his own opinions rather than scientific evidence. Weed management is complex and dependent on scale and local conditions as well as local regulations; it would have been useful if this had been discussed in some detail. Moreover, the treatments of the various control methods are cursory, especially considering the volume of information available in the literature. I would have preferred to see an actual literature review of control methods for aquatic systems with appropriate citations, so the reader could consider other viewpoints. However, a list of websites, including government websites, and references are included at the end of the book and provides the reader with the opportunity to do this. The section on assessment and planning raises some important points and provides a useful list of key issues that should be included in any management program. An important point that was not covered related to the importance of understanding the local ecology of the weeds. In certain circumstances removal of weeds can be harmful to fauna that depend on them (Carlos and Gibson 2010; Jayawardana *et al.* 2010), or may cause erosion of banks (Zukowski and Gawne 2006). Thus, weed removal should be carried out in gradual stages in conjunction with planting of natives, to replace the environmental services provided by the weeds.

In Chapter 3 Romanowski discusses how Australian native plants can, and have, become weeds, an important topic little recognised by the public. He then proceeds to discuss the origins, uses, preferred growth conditions, species that can be confused with each other, environmental impacts and values and control and management of minor indigenous wetland weeds. Species are discussed under the genus in which they occur. This is disconcerting as not all species within these genera are weeds. Chapter 4 presents a compendium of weeds and largely follows the format used for minor indigenous wetland weeds in Chapter 3. In Chapter 4, however, the weeds are firstly divided into: grasses; sedges, rushes and other relatives of grasses; other wetland weeds; hardy waterlilies, tropical waterlilies; algae and cyanobacteria and seaweeds. These two chapters are useful and provide the reader with a good idea of what the problem plants are and for which

species they should be on the alert.

Thirty-two coloured plates are included and depict various weedscapes, highlighting the invasive nature of many of the species pictured. Other photographs are useful identification aids. The photographic plates are grouped together between pages 30 and 31 but I am sure many readers would prefer a coloured photograph of each species in the compendium, alongside their associated information. The glossary provides informal definitions of more unusual terms and would be useful to those unfamiliar with such terms. I feel the book is a little overpriced but it would make a useful addition to the library of those who care for our wetland environments. It would facilitate their recognition of what plants to remove when restoring a wetland, which to use in revegetation of a wetland or creation of a new wetland, even if this wetland is only a small pond in the backyard.

References
Adair RJ and Groves RH (1998) Impact of environmental weeds on biodiversity: a review and development of a methodology. Occasional Publication, National Weeds Program, Environment Australia, Canberra.
Carlos EH and Gibson M (2010) The habitat value of Gorse *Ulex europaeus* L. and Hawthorn *Crataegus monogyna* Jacq. For birds in Quarry Hills Bushland Park, Victoria. *The Victorian Naturalist* 127, 115-124.
Gorham P (2008) Aquatic weed management in waterways and dams. Primefacts profitable and sustainable primary industries. Primefact 30. http://www.dpi.nsw.gov.au/__data/assets/pdf_file/0020/256403/Aquatic-weed-management-in-waterways-and-dams.pdf accessed 5 September 2012.
Groves RH, Boden R and Lonsdale WM (2005) Jumping the Garden Fence: Invasive garden plants in Australia and their environmental and agricultural impacts. A CSIRO report for WWF Australia.
Jayawardana JMCK, Westbrooke M and Wilson M (2010) Leaf litter decomposition and utilisation by macroinvertebrates in a central Victorian river in Australia. *The Victorian Naturalist* 127, 104-114.
Moran P (2009) Aquatic weeds.......so what? http://moosabiosphere.org.au/_blog/Environment_Blog/post/aquatic_weedsso_what accessed 5 September 2012.
Zukowski S and Gawne B (2006) Potential effects of Willow (*Salix* spp.) removal on freshwater ecosystem dynamics: a literature review. Report for the North East Catchment Management Authority. Murray-Darling Freshwater Research Centre, Wodonga.

Maria Gibson
Environmental Sustainability Research Group
Deakin University
221 Burwood Highway
Burwood, Victoria 3125

Kangaroos

by Terence Dawson

Publisher: *CSIRO publishing, Collingwood, 2012. 216 pages, paperback. ISBN 9780643106253. RRP $ 39.95*

In 1996 in an editorial in the esteemed journal *Conservation Biology* the then editor Reed Noss wrote an article provocatively titled 'Are the Naturalists Dying Off?', in which he decried the loss of natural history skills in modern day conservation biologists. He wrote how important natural history was in providing the rigorous, first-hand observations that underpinned conservation biology and provided the data for predictive models, hypothesis posing and even enthusiasm for the newly emerging discipline.

The *Australian Natural History Series* (initially published by UNSW Press but now by CSIRO Publishing) is an excellent series that has over the years provided that wonderful nexus between natural history and science—a science that is challenged to account for astute observations made in the field. *Kangaroos*, now in its second edition, is an exemplar in the series and written by one of Australia's leaders in the field of marsupial biology, Terry Dawson.

A good example of this interrelationship between natural history and science is from my own field of marsupial thermoregulation. Kangaroos and wallabies have been observed to lie in the heat with their naked areas of skin exposed; further, they lick their fur and pant like dogs. Conventional wisdom explained these observations by (correctly) stating heat loss is enhanced by peripheral vasodilatation of the blood vessels in the unfurred skin—the process having the quaint descriptor of 'opening up the thermal windows'; panting and fur licking are examples of evaporative cooling.

Kangaroos at rest do not sweat, yet microscopic examination of their skin reveals advanced sweat glands - the so called eccrine glands - which produce sweat in other animals. Furthermore, kangaroo hunters have described how animals that have been chased are often covered in sweat.

Dawson's laboratory has undertaken experiments to demonstrate that kangaroos do indeed sweat, but only when exercising. As well, his team has shown that fur licking was not a primitive adaptive response to heat but instead, an advanced one in which saliva is smeared over the wrist areas which have superficial blood vessels; the evaporation thus cools the blood very effectively.

These examples illustrate the underlying theme of the book—observations and the underpinning science that offers explanations. After introductory chapters on evolution and diversity of kangaroos, there are chapters on population structure, social organisation, reproduction, feeding, water and temperature regulation and finally chapters on human dimensions of kangaroo interactions and management.

These were the same chapters as in the first

edition. So what has been changed? The first major addition has been the inclusion of more data to substantiate and illustrate the biology of kangaroos. For example, we now have a graph showing that kangaroos have an amazing ability to increase speed of hopping with little increase in metabolic rate (and thus energy expenditure). Methods of age estimation of animals are provided in some detail. There are some other minor improvements: there is now one consolidated reference list instead of listing references per chapter.

Secondly, the material has been significantly updated. I have reviewed some new editions of books that have barely changed; this is not the case with *Kangaroos*. Indeed, I counted some 47 new references published since the original edition. The book is written well and is illus-trated to good effect.

Who is the audience for this book? I would expect naturalists who ponder on the significance and adaptive advantages of behaviour, physiology and anatomy would enjoy the book. Scientists will appreciate its rigorous, evidence-based approach. It is not a coffee table book (although there are 16 colour plates); rather, it is a book that seeks to demonstrate and explain the remarkable, advanced adaptations kangaroos have to their many environments.

Rob Wallis
Horsham Campus Research Precinct
University of Ballarat
Horsham. Victoria 3402

Australia's Amazing Kangaroos: Their conservation, unique biology and coexistence with humans

by Ken Richardson

Publisher: *CSIRO publishing, Collingwood, 2012. 240 pages, paperback.
ISBN: 9780643097391. RRP $ 49.95*

Some years ago at an international conference I had difficulty in explaining to some delegates from where I came. Some thought I came from Austria, while others were convinced I came from the UK. The problem was solved by my hopping briefly with arms in front—aha, Australia, they all said at once. Kangaroos = Australian!

Ken Richardson's excellent book aims to 'bridge the ever-widening gap between the mountains of detailed information found in the serious scientific literature and the many members of the public who wish to be better informed about Australia's iconic kangaroos. In today's rapidly changing world, the better we are informed about our native animals the better their prospects for survival' (page v).

The result is a scientifically authoritative, contemporary, beautifully illustrated book that is

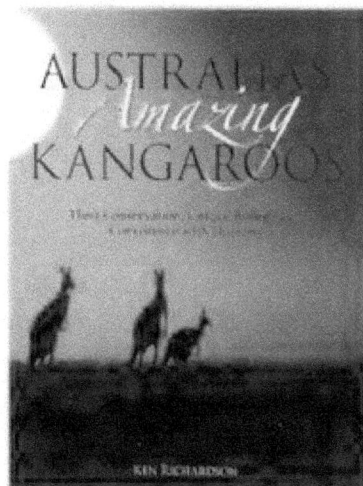

essentially written in two sections.

In the first, a brief account of marsupial evolution, kangaroo characteristics and conservation status precedes a species by species account of each member of the suborder Macropodiformes. This section could be considered an update on Ron Strahan's (and Steve Van Dyck's updated) seminal work *The Mammals of Australia* – complete with the species' distinctive features, distribution, threatening factors, management actions and an excellent photograph.

The second section covers Adaptation and Function (morphological adaptations, mobility and movement, diet, reproduction etc.), Conservation, and finally Kangaroos and Humans Today, which includes an informative account of the kangaroo harvesting industry.

This last chapter distinguishes this book from many others on today's market, including Dawson's *Kangaroos*, which I have reviewed previously in this issue (see p. 56). For example, a most useful Appendix outlines procedures to be used when kangaroos are commercially harvested from the four states in which the industry is legally operating. Hopefully this chapter will inform the current and sometimes heated debate about harvesting kangaroos as a useful resource versus protecting a much-loved natural icon. Of course, such a discussion is fraught with challenges, with protagonists arguing from quite different platforms (ethical/moral/aesthetic versus pragmatic/resource utilisation).

This is a well written and well presented book. Its acknowledgements are a who's who of Australian mammalogy and the photographs are excellent.

So who will gain most by buying the book? Readers will need a reasonable background knowledge of mammalian biology. They will need an understanding of biological terms, although there is a glossary to help a reader's comprehension. The book will certainly appeal to naturalists – the first section will help in macropodid identification while the latter chapters provide anatomical, physiological and behavioural bases for observations people make in the field.

Biology students will enjoy the clear presentation, up to date science and lucid explanation of some quite sophisticated biological concepts. The case studies—Heirisson Prong Peninsula (Shark Bay) reintroduction of the Burrowing Bettong and the Proserpine Rock Wallaby recovery plan—provide excellent accounts of conservation management with realistic evaluations of their success.

In the interests of readability, sources are not cited in the text and the bibliography is thin and quite general. This detracts from its usefulness to researchers and students alike. Steve Van Dyck's name is misspelled and Emeritus Professor Ian Hume is given two such awards, but these are minor distractions from an otherwise most valuable addition to the popular literature on what most people regard as the most Australian of animals — the kangaroos.

Rob Wallis
Horsham Campus Research Precinct
University of Ballarat
Horsham, Victoria 3402

A Natural History of Australian Bats - Working the Night Shift

by Greg Richards and Les Hall with photography by Steve Parish

Publisher: *CSIRO Publishing, Collingwood, 2012. viii, 192 pages, hardback, colour photographs. ISBN 9780643103740. RRP $79.95*

A Natural History of Australian Bats - Working the Night Shift by Greg Richards and Les Hall is an absolutely tremendous book, which introduces the wonderful world of bats. The authors, who have worked for over 40 years on bats (as you can tell), present a broad range of topics on Australian bats in a captivating and descriptive way.

With over 400 large beautiful colour photographs, mostly by the acclaimed photographer Steve Parish, it is written for the general public, naturalists and students. But I am sure that scientists will also find it enjoyable, and learn a few things, as I certainly did.

The page size is large—A4—and so is the text, which is succinct and clear. The text of the book is 184 pages long, and condenses major topics on our current knowledge about Australian bats into eight punchy chapters.

The first chapter opens by cultivating a general fascination about bats, and you can easily see the passion of the authors for these nocturnal mammals. A short overview of significant events in bat research history follows, including the development of specialised research equipment.

The next chapter, the Travelogue, presents some of the characteristic species, as well as important bat habitats and noteworthy locations for bats from each 'bat bioregion', from significant islands, down to the major cities in Australia. This part might be especially enjoyable for grey nomads (bat veteran researchers or enthusiasts). More specific details are then presented in chapters 3 to 5 on the sophisticated and intriguing bat morphology, bat breeding behaviour and general ecology. These chapters constitute nearly one third of the book and provide a great overview as well as interesting facts on major topics in these areas.

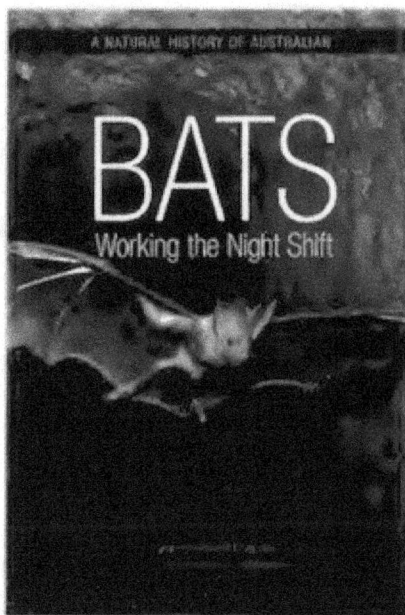

A disturbing part of the book is chapter 6, where trials and tribulations of being a bat in Australia are described. Short and diverse paragraphs cover a range of issues, from the natural predators of bats to the serious impact of humans and the potential future impacts of global warming, clearly showing the risks and threats bats are exposed to. Chapter 7, which refers to the book title, starts by describing the fossil history of bats. It then gives an overview on bats in the culture of Aboriginal Australians, in prehistoric paintings in Australian caves, and goes on to detail the first encounter of European explorers with bats in Australia. Also highlighted are the passionate people who rescue and care for injured bats. The closing chapter is devoted to species profiles, with stunning photos of a large proportion of Australian bats, and associated descriptive information.

What I liked especially about this book is that it brings together current knowledge on bats in Australia, and each chapter is presented in a way that can be followed easily even by people who are completely new to the world of bats. In addition, it manages to present factual information that would intrigue bat scientists, making this an enjoyable read for them too. All readers benefit greatly by the profusion of photographs that enhance the text to make this book a very engaging read. I guarantee that, by the end of this book, readers will be hooked on bats, if they are not hooked already!

Tanja Straka
Australian Research Centre for
Urban Ecology (ARCUE)
c/o School of Botany
The University of Melbourne

A Guide to Australia's Spiny Freshwater Crayfish.

By Robert B McCormack

Publisher: *CSIRO Publishing, Collingwood, Vic., Paperback 2012. 248 pages, paperback colour photographs. ISBN: 9780643103863. RRP $59.95.*

From time to time I have been asked to identify freshwater crayfish for environmental managers, curious members of the public or kids. While I know something about marine crustaceans, these groups are not my speciality. So a 'guide' is just what I need. This new book is a fine publication but it is not a guide for those who might want to find out the name of a newly caught crayfish. Identification of species, especially in a genus like *Euastacus* with 50 named species and more yet to be described, is no easy task. To the uninitiated (that's most of us) telling one from the other is difficult. Gary Morgan, whose taxonomic work (1986–1997) is the foundation of current understanding, provided dichotomous keys full of arcane terms and demanded an appreciation of subtle distinctions. No substitute for these keys is provided in this book—perhaps that is not possible but some of the new information provided here, colour patterns for example, might have proved useful. I wonder how the author and his colleagues identify species without resorting to Morgan's

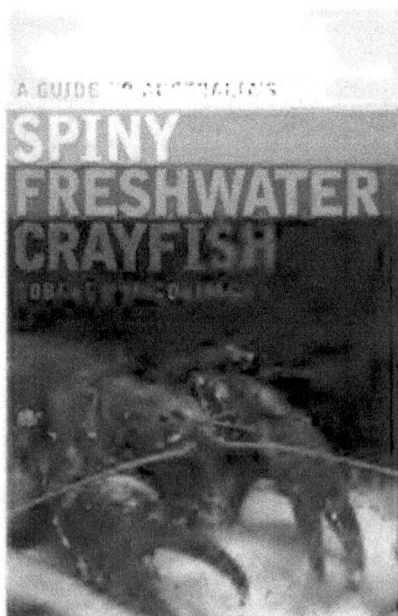

works. The section devoted to identification of crayfish relies on division of the species into what the author calls 'giant', 'intermediate' and 'dwarf' spiny crayfish, each group characterised by about 20 morphological, ecological and behavioural features, some overlapping. I found it difficult to differentiate some of the species tabulated aphabetically on the basis of these characters. The section fails to explain how to tell one species from another.

Having said that, *A Guide to Australia's Spiny Freshwater Crayfish* should have general interest. I am a fan of books that cover a well defined taxonomic group, the genus *Euastacus* in this case, in detail and in a popular format. Three-quarters of the book discusses all 50 species found in Australia in alphabetical order (which is fine if you know the species name of the species of interest). A colour photograph, ecology, diagnostic features, distribution, colour, size, breeding and conservation status are given for each. We learn that many are uncommon and most are restricted to small catchment areas. In the absence of identification keys, it might have been better to group similar species together and discuss how they differed from their neighbours. The introductory chapters discuss basic crayfish anatomy, moulting and growth, morphology and breeding. A concluding chapter evaluates threats from recreational fishing, illegal fishing, climate change, habitat alteration, exotic species and diseases.

There is a wealth of information in the introductory chapters and for each species. I assume that the author and his colleagues in the Australian Crayfish Project, a collaboration active since 2005, are responsible for most of it and are to be commended for bringing it together. I was curious why only 27, mostly taxonomic, citations are listed in the *References* section. A vast number of refereed papers and government reports goes unacknowledged. This literature deals with spiny crayfish physiology, reproduction, life history, evolution, phylogeny, population genetics, conservation, management, fisheries, diet, behaviour, growth, burrows, Aboriginal use, moulting, pollution tolerance, dispersal, ectoparasites and more (try *Euasta-*

cus in Google Scholar). I list below some key papers from the hundreds I found.

The writing style ranges from highly technical to chatty, fact and opinion often intermingled. The technical sections will be beyond many readers – I wonder what are the really diagnostic bits in the species diagnoses. Advice on handling spiny crayfish is useful but not in a section on anatomy (p. 32). The excellent coloured and labelled illustrations of anatomy (pp. 37-40) seem to stand alone and the difference between 'small', 'medium' and 'large' spines is never explained. Small errors have crept in. The cheliped (and in fact all the legs) comprise seven segments, coxa, basis, ischium, merus, carpus, propodus and dactylus in that order, not four as stated. There is only one pair of uropods, each with two branches, not two pairs.

This book covers a lot and fills a gap for an important and charismatic group of endemic Australian crustaceans and for these reasons it is to be recommended.

References
Crandall, KA, Fetzner JW, Lawler SH, Kinnersley M and Austin CM (1999) Phylogenetic relationships among the Australian and New Zealand genera of freshwater crayfishes (Decapoda: Parastacidae). *Australian Journal of Zoology* 47, 199-214.
Furse JM and Coughran J (2010) An assessment of genus *Euastacus* (49 species) versus IUCN Red List criteria. A report prepared for the Global Species Conservation Assessment of lobsters and freshwater crayfish for the IUCN Red List of Threatened Species. International Association of Astacology.
Horwitz P (1995) *A preliminary key to the species of Decapoda (Crustacea: Malacostraca) found in Australian inland waters* (Co-operative Research Centre for Freshwater Ecology Identification Guide 5). (Co-operative Research Centre for Freshwater Ecology: Albury, NSW)
Zukowski S, Watts R and Curtis A (2012) Linking biology to fishing regulations: Australia's Murray Crayfish (*Euastacus armatus*). *Ecological Management & Restoration* 13, 183-190.

Gary CB Poore
Museum Victoria
GPO Box 666
Melbourne 3001

Australian High Country Owls

by Jerry Olsen

Publisher: *CSIRO Publishing, Collingwood, 2012. 376 pages, paperback, colour photographs.*
ISBN: 9780643097056. RRP $69.95

Owls have an uncanny ability to intrigue us, whether through their regular nightly rituals, particularly during their breeding season, their persistent calling in the depths of the night, their elusiveness (causing subsequent frustration on our part while trying to find them) or simply because of their mystery and awe. Owls, for whatever reason, hold a very special place in the hearts of many people.

As public awareness about owls increases, so do the questions about these cryptic creatures. Jerry Olsen, through his book *Australian High Country Owls*, has provided extensive information and knowledge on both Australian and international owl species. The 366 page book provides detailed accounts from Jerry's own experiences as well as incorporating relevant scientific literature. The book focuses primarily on Australian *Ninox* species and how these compare to their international counterparts. This comparison is extremely valuable, as highlighted in Jerry's preface, because much of the owl research undertaken in Australia is based on protocols and data interpretation from overseas.

The book contains 41 chapters, all of which are very well written and easy to read. The first few chapters contain essential background information such as what an owl is, owl taxonomy, in particular *Ninox* species followed by the Southern Boobook as a *Ninox* example. After providing background information such as this, the book delves into the many aspects of studying owls, including how to locate, trap and handle these often difficult creatures on which to undertake research!

Detailed information and accounts of diet and hunting, breeding and conservation are also provided. These sections are extremely valuable as they draw heavily on both the scientific literature and the author's own research experiences. Summary tables, figures and photographs are well used to provide information and highlight visual points. The photography is of extremely high quality and certainly provides further insights into the mystery associated with these birds.

The final section of the book is titled Wallacea and provides a detailed account of owls on the island of Sumba and the discovery of a new owl species, the Little Sumba Hawk-owl. An extensive reference list is also provided along with appendices on Australian owls and rehabilitating injured or orphaned owls.

Jerry Olsen is one of Australia's leading owl researchers and has many years of experience working with these birds. This expertise and experience shines throughout this book, with interesting and accurate information. The book is a must have for anyone interested in learning more about these mysterious and amazing creatures.

Dr Raylene Cooke
School of Life and Environmental Sciences
Deakin University – Melbourne Campus
221 Burwood Highway
Burwood Vic 3133

Australian Natural History Medallion Trust Fund

Donations to the Trust Fund were gratefully received during 2012 from the following:

Julia Davis	$ 10.00
Helen Handreck	$ 10.00
David Cheal	$ 5.00
Elizabeth Sevior	$ 40.00
Latrobe Valley FNC	$ 50.00
Portland FNC	$ 50.00
Helen Aston	$100.00
WA Naturalists Club Inc	$100.00
John Poppins	$ 20.00
Valda Dedman	$ 10.00
Launceston Field Nats	$ 50.00
Alan Reid	$ 16.00
Ken Simpson	$ 50.00
Maryborough FNC	$ 50.00
Kay Taranto	$ 10.00

If you would like to contribute to this fund, which supports the Australian Natural History Medallion, donations should be sent to: The Treasurer, Field Naturalists Club of Victoria, Locked Bag 3, Blackburn, Vic. 3130. Cheques should be made payable to the 'Australian Natural History Medallion Trust Fund'.

The medallion is awarded annually to a person who is considered to have made the most significant contribution to the understanding of Australian natural history in the last ten years.

Gary Presland
Secretary
Australian Natural History Medallion

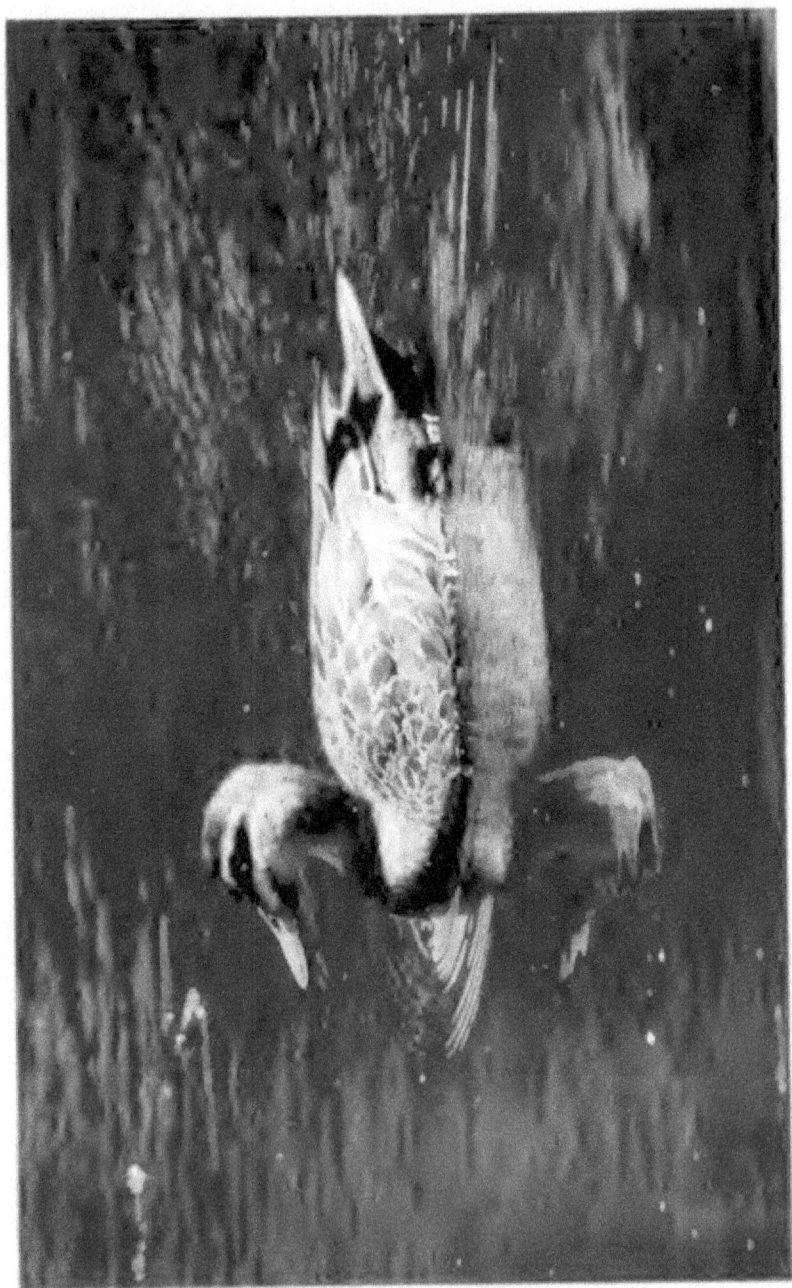

www.ingramcontent.com/pod-product-compliance
Lightning Source LLC
Chambersburg PA
CBHW022009190326
41519CB00010B/1446